CHARLES DARWIN IN CAMBRIDGE

The Most Joyful Years

"Upon the whole, the three years I spent at Cambridge were the most joyful of my happy life."

CHARLES DARWIN IN CAMBRIDGE

The Most Joyful Years

John van Wyhe

NEW JERSEY • LONDON • SINGAPORE • BEIJING • SHANGHAI • HONG KONG • TAIPEI • CHENNAI

Published by

World Scientific Publishing Co. Pte. Ltd.

5 Toh Tuck Link, Singapore 596224

USA office: 27 Warren Street, Suite 401-402, Hackensack, NJ 07601

UK office: 57 Shelton Street, Covent Garden, London WC2H 9HE

Library of Congress Cataloging-in-Publication Data
Van Wyhe, John, 1971–
Charles Darwin in Cambridge : the most joyful years / John van Wyhe, National University of Singapore, Singapore.
pages cm
Includes bibliographical references and index.
ISBN 978-9814583961 (hardcover : alk. paper) -- ISBN 978-9814583978 (softcover : alk. paper)
1. Darwin, Charles, 1809–1882--Knowledge and learning. 2. Darwin, Charles, 1809–1882--Homes and haunts. 3. Darwin, Charles, 1809–1882--Childhood and youth. 4. Naturalists--England--Biography. 5. University of Cambridge. I. Title. II. Title: Darwin in Cambridge.
QH31.D2V355 2014
576.8'2092--dc23
[B]

2013046755

British Library Cataloguing-in-Publication Data
A catalogue record for this book is available from the British Library.

Cover Image: Engraving of Christ's College, by J. Le Keux after I. A. Bell. Published April 1838.

In-house Editor: Darilyn Yap

Typeset by Stallion Press
Email: enquiries@stallionpress.com

Printed in Singapore by FuIsland Offset Printing (S) Pte Ltd

In memory of Professor Malcolm Bowie FBA (1943–2007),
Master of Christ's College 2002–2006.

Chalk and water-colour drawing of Charles Darwin in 1840
by George Richmond. Reproduced courtesy of the Darwin Heirlooms Trust.

Contents

Introduction

Charles Darwin's years as a student at the University of Cambridge were some of the most important and formative of his life. Thereafter he always felt a particular affection for Cambridge. For a time, he even considered a Cambridge professorship as a career and he sent three of his sons there to be educated. Unfortunately the remaining traces of what Darwin actually did and experienced in Cambridge have long remained undiscovered. Consequently his day-to-day life there has remained unknown and misunderstood. This book is based on new research, including newly discovered manuscripts and Darwin publications, and gathers together recollections of many of those who knew Darwin as a student. Darwin's time in Cambridge is therefore revealed here in unprecedented detail.

This book is based on a booklet I wrote for the 2009 bicentenary commemorations of Darwin while I was a Bye-Fellow of Christ's College, Cambridge. That work was printed in 2000 copies which sold out by the end of the year. Since that time I have continued to receive letters asking me where it could be obtained and the price for second-hand copies on the internet has reached $195. This gratifying and continuing interest in the book and a number of new research findings and additional materials which I wanted to add to the earlier treatment but which could not fit into the original format have led me to produce the present work.

As a historian of science specializing on Darwin and the director of the University of Cambridge project *The Complete Work of Charles Darwin Online*, the privilege of experiencing and being part of the life of Darwin's own college, Christ's, was enormously interesting and enriching. I owe this opportunity primarily to the interest and kindness of the late Master Malcolm Bowie.

During my years at Christ's, I could not resist searching for any traces of Darwin that remained, as yet unknown to historians. For example, there was an old framed photograph of Darwin in the Old Library with a signature displayed through the mount underneath. Was this the signature cut from a letter? I requested that the photograph be taken down and examined. The signature turned out to be the endorsement on the back of an intact cheque from 1872. This little discovery generated a small amount of media interest, surely highlighted by the British tabloid newspaper *The Sun*, which quipped "Cheque this out Darwin" although the by-line of the *Cambridge Evening News* was also noteworthy "Academic's Darwin hunch cheques out." The story of the cheque, photograph and how they all came to hang in the library at Christ's is told below.

I frequently urged and, I suppose, plagued my dear friend and the Honorary Keeper of the Archives and Fellow Commoner Geoffrey Martin to search the archives again for any traces of Darwin. For a hundred years, records of Darwin had occassionally been sought at Christ's. All that was known was his name in the admissions book. Geoffrey however made wonderful discoveries in the form of a series of large leather-bound record books which contained Darwin's College bills. With the permission of the College, these were published on *Darwin Online* to considerable international media attention. These and other discoveries made it possible to give a new account of Darwin in Cambridge focused more on student life within the university as he experienced it. This book provides a wealth of information not previously available in any other work and corrects several errors in the literature on this pivotal chapter in Darwin's life. The text has been amended, extended and many new illustrations are added. For more on Darwin's life in these years consult the first volume of *The Correspondence of Charles Darwin, the Life and letters* vol. 1, pp. 163–184, Janet Browne, *Voyaging*, Desmond and Moore, *Darwin*, Tony Larkum, *A natural calling* and Keith Thomson, *Young Charles Darwin*.

John van Wyhe
Singapore
September 2013

EARLY LIFE 1809–1825

Charles Robert Darwin was born on 12 February 1809, the fifth of six children of a wealthy family in the small market town of Shrewsbury, Shropshire.[1] 1809 was the period now remembered as the setting for Jane Austen's novels and the Napoleonic Wars in which Britain was the primary opponent for supremacy in Europe and far beyond. Napoleon was finally defeated in 1815. The post war era would see Britain become the world's greatest military and economic power for over a century. Another consequence of the wars was that Spain lost her South American colonies. With the end of Spain's control, there was an opening up of trading opportunities with her former colonies. In the 1820s Britain would seek to strengthen ties there and thus sent out naval surveying and mapping expeditions, two of which were by a small ship called HMS *Beagle*. This, very briefly stated, is the world into which Darwin was born.

His father, the portly financier and physician Robert Waring Darwin (1766–1848), was the son of the noted poet and physician Erasmus Darwin (1731–1802). Darwin's mother, Susannah Wedgwood (1765–1817), was the daughter of Josiah Wedgwood I, the well-known industrial potter. Despite the fact that his father was a freethinker and his mother a Unitarian, Darwin was baptized in the Parish Church of St. Chad's on 15 November 1809. The Parish Register of Christenings

[1] For an accessable and illustrated biography of Darwin see Wyhe, John van. 2008. *Darwin*. London: Andre Deutsch and New York: National Geographic [2009].

and Burials entry reads: "Darwin Chas. Robt. Son of Dr. Robt. & M^{rs}. Susannah his wife/born Feb^{r}. 12th". Darwin's official membership of the Church of England would open up forms of social advancement that were closed to those not members of the established church such as attendance at an English University. Darwin grew up in a large comfortable house with many servants and many opportunities.

He was first tutored at home before attending a day-school in Shrewsbury run by the Rev. George Case, minister of the local Unitarian Chapel. Darwin later recalled:

> By the time I went to this day-school my taste for natural history, and more especially for collecting, was well developed. I tried to make out the names of plants, and collected all sorts of things, shells, seals, franks, coins, and minerals. The passion for collecting, which leads a man to be a systematic naturalist, a virtuoso or a miser, was very strong in me, and was clearly innate, as none of my sisters or brother ever had this taste.[2]

Charles, aged 6, and Caroline Darwin, aged 5, c. 1815 by Ellen Sharples.
Reproduced courtesy of the Darwin Heirlooms Trust.

[2] *Autobiography*, p. 22. It has sometimes been claimed that Darwin never referred to this manuscript, titled 'Recollections of the Development of my mind & character' as his autobiography. However in his 'Journal' and his last will and testament, published for the first time on *Darwin Online*, Darwin refers to this document as 'my Autobiography'. See the new transcription by Kees Rookmaaker of the *Autobiography* side-by-side with facsimilies of the manuscript on *Darwin Online* at http://darwin-online.org.uk/content/frameset?viewtype=side&itemID=CUL-DAR26.1-121&pageseq=1.

Coming from a long line of liberal Unitarians, Darwin's mother took her children to the Rev. Case's chapel on Sundays. Darwin's mother died in July 1817 when he was only eight years old. It used to be claimed that her death profoundly affected him, particularly when Freudian-inspired theories were popular. However, there is almost no evidence of any feelings about her of any kind. In a household where he was raised and cared for by elder sisters and maidservants, his mother's loss at such an early age was not as grave as it might otherwise seem. For his elder siblings, the blow was far more profound.

From 1818 to 1825 Darwin was a pupil at the Free Grammar School at Shrewsbury, about a mile from the family home. It was said to be one of the best schools in England and was run by Dr Samuel Butler. Darwin later recalled:

> I remember in the early part of my school life that I often had to run very quickly to be in time, and from being a fleet runner was generally successful; but when in doubt I prayed earnestly to God to help me, and I well remember that I attributed my success to the prayers and not to my quick running, and marvelled how generally I was aided.[3]

Shrewsbury Free Grammar School. The building today houses the public library.

Darwin later felt that his time at school was wasted learning Greek and Latin. He studied chemistry in a home laboratory in a garden "tool-house" with his elder brother Erasmus Alvey Darwin (1804–1881), usually called Ras or Eras

[3] *Autobiography*, p. 25. Details of the history of the school and its curriculum can be found in Carlisle, N. 1818. *A concise description of the endowed Grammar Schools in England*, vol. 2. London: Baldwin, Cradock and Joy. See Pattison, Andrew. 2009. *The Darwins of Shrewsbury*. Stroud: History Press.

by the family. The lessons he learned through carefully studying chemistry textbooks and carrying out experiments were important childhood training. He also loved reading, fishing and solitary walks.

> [The chemistry] was the best part of my education at school, for it showed me practically the meaning of experimental science. The fact that we worked at chemistry somehow got known at school, and as it was an unprecedented fact, I was nick-named 'Gas.' I was also once publicly rebuked by the head-master, Dr. Butler, for thus wasting my time over such useless subjects[4]

Perhaps just as beloved by Darwin was his passion for shooting partridges, pigeons, rabbits and rats. An interest in natural history and systematic record keeping were already evident. He loved learning about the habits and habitats of his prey, and kept an exact record of his victims by knotting a piece of string tied to a button hole. He later recalled:

> One day when shooting [with friends] I thought myself shamefully used, for every time after I had fired and thought that I had killed a bird, one of the two acted as if loading his gun and cried out, "You must not count that bird, for I fired at the same time," and the gamekeeper perceiving the joke, backed them up. After some hours they told me the joke, but it was no joke to me for I had shot a large number of birds, but did not know how many, and could not add them to my list...[5]

---

[4] *Autobiography*, p. 46.
[5] *Autobiography*, p. 54.

EDINBURGH UNIVERSITY 1825–1827

In October 1825 Darwin's father sent him, aged only 16, to the University of Edinburgh to study medicine along with his brother Erasmus. Erasmus had been admitted a student at Christ's College, Cambridge on 9 February 1822. He was continuing his medical studies in Edinburgh which was far ahead of Cambridge in medicine. In going to Edinburgh, Darwin was following in the tradition of his father and grandfather. The two Darwins lodged together at 11 Lothian Street. The instruction at Edinburgh was through lectures and demonstrations. Darwin found the lectures "intolerably dull". Out of his own interest, he also attended lectures on natural science. He praised most highly Thomas Hope's entertaining chemistry lectures. Darwin took copious notes which still survive.[6] He also attended lectures on geology by Robert Jameson, the Regius Professor of Natural History. Darwin later claimed that Jameson's old-fashioned geology and ungentlemanly sneers at the more recent Plutonian or volcanic theories of colleagues convinced Darwin, for a time, to ignore geology.[7] Edinburgh University

[6] Some of these notes are reproduced in Wyhe, John van. 2008. *Darwin*. London: Andre Deutsch and New York: National Geographic [2009], and all of them are included in Wyhe, John van. (ed.) 2002. *The Complete Work of Charles Darwin Online* (http://darwin-online.org.uk).

[7] Secord, James. 1991. The discovery of a vocation: Darwin's early geology. *British Journal for the History of Science* 24: 133–157 and Herbert, Sandra. 2005. *Charles Darwin, geologist*. Ithaca, N.Y: Cornell University Press.

was undoubtedly an important formative influence on Darwin. It was markedly secular compared to the English universities and he learned important lessons about contemporary science which he retained throughout his life.

The University of Edinburgh. Engraving by Thomas Shepherd 1829/30.

Darwin later recalled: "My Brother staid only one year at the University, so that during the second year I was left to my own resources; and this was an advantage, for I became well acquainted with several young men fond of natural science."[8] The most influential of these young men was Dr Robert Edmond Grant (1793–1874), an expert on marine invertebrates and a devotee of the transmutation theory of the great French naturalist Jean-Baptiste Lamarck. Darwin accompanied Grant on collecting trips in the Firth of Forth and became very interested in the study of marine invertebrates, an interest he would maintain for a quarter of a century.[9] "I also became friends with some of the Newhaven fishermen," Darwin recalled, "and sometimes accompanied them when they trawled for oysters, and thus got many specimens." Dissecting and observing his captures under a "wretched microscope", Darwin made several new observations. These he reported to the student Plinian Society although the papers were not published. Darwin seems also to have briefly flirted with the unorthodox character science of phrenology which was at just this time at its intellectual height. Edinburgh and its medical community were the national centre for the growing phrenology

[8] *Autobiography*, p. 48.

[9] See Secord, James. 1991. Edinburgh Lamarckians: Robert Jameson and Robert E. Grant, *Journal of the History of Biology* 24: 1–18.

'movement'.[10] Darwin paid a former black slave from Guiana named John Edmonstone for bird stuffing lessons. This was essential to preserving specimens and would prove of great service during the *Beagle* voyage.

While at Edinburgh Darwin discerned that his father would leave him enough property to live in comfort, thus dispelling any real sense of urgency in learning medicine, which Darwin disliked. Most of all he was unable to stand the sight of blood or suffering, traits that he retained throughout his life.

Darwin left Edinburgh in April 1827 and toured Dundee, St Andrew's, Stirling, Glasgow, Belfast and Dublin, his only visit to Ireland. At the end of May, he made his only visit to the continent when he travelled with his sister Caroline and Josiah Wedgwood [II]. Throughout the autumn he continued his visits and pursued his favourite hobby of shooting, especially in the nearby estate of Woodhouse, the home of William Mostyn Owen and his pretty daughters. The Owens and the Darwins were good friends. Darwin and the second daughter Frances "Fanny" were soon flirting playfully during these visits.

Darwin's father finally accepted that Darwin had no desire or suitability to become a physician. He proposed instead that Darwin become a clergyman:

> He was very properly vehement against my turning an idle sporting man, which then seemed my probable destination. I asked for some time to consider, as from what little I had heard and thought on the subject I had scruples about declaring my belief in all the dogmas of the Church of England; though otherwise I liked the thought of being a country clergyman. Accordingly I read with care *Pearson on the Creed* and a few other books on divinity; and as I did not then in the least doubt the strict and literal truth of every word in the Bible, I soon persuaded myself that our Creed must be fully accepted.[11]

Darwin also read the Revd John Bird Sumner's, *The evidence of Christianity, derived from its nature and reception* (1824).[12] His reading notes still survive. Sumner's arguments seemed to show, and Darwin's notes reveal that he was persuaded, that Jesus must have lived and been the son of God.

---

[10] See Wyhe, John van. 2004. *Phrenology and the origins of Victorian scientific naturalism*. Ashgate.

[11] *Autobiography*, p. 58.

[12] The notes were first transcribed and published as Darwin, C.R [c. 1827.] [Notes on reading Sumner's *Evidence of Christianity*]. CUL-DAR91.114–118, transcribed by John van Wyhe. *Darwin Online*, http://darwin-online.org.uk/.

> As it was decided that I should be a clergyman, it was necessary that I should go to one of the English universities and take a degree; but as I had never opened a classical book since leaving school, I found to my dismay that in the two intervening years I had actually forgotten, incredible as it may appear, almost everything which I had learnt even to some few of the Greek letters. I did not therefore proceed to Cambridge at the usual time in October, but worked with a private tutor in Shrewsbury and went to Cambridge after the Christmas vacation, early in 1828. I soon recovered my school standard of knowledge, and could translate easy Greek books, such as Homer and the Greek Testament with moderate facility.[13]

Previous biographies never explained why Darwin needed to catch up on his Greek before entering Cambridge. A more detailed account of undergraduate Cambridge life makes it clear that it was essential to be proficient in Greek and Latin because the daily college lectures and private tutorials were conducted at an advanced level. I once showed a Christ's College examination paper from Darwin's time to a Cambridge professor of Classics who remarked to me: "I couldn't do this!"

For Darwin, who was never very religious, becoming a clergyman probably meant a secure and undemanding career for someone of his genteel social background. After all, he knew plenty of country parsons amongst his friends and family. And he would be left with plenty of time to pursue his natural history interests like the famous parson naturalist Gilbert White (1720–1793), author of the remarkably popular *Natural History of Selborne* (1789).

[13] *Autobiography*, p. 58.

COMING UP TO CAMBRIDGE

Cambridge in 1828 was a quiet market town on the broad flat plains of fenland surrounded by reclaimed and many still unreclaimed marshes of East Anglia. The town was bisected by the meandering river Cam. The town, with a population of 21,000, was about one and a half miles long from north to south and one mile wide. The university population, all male, was about 2,000. It was said that Oxford was a university in a town and that Cambridge was a town in a university. In the centre of the town was the marketplace. The food and goods available here were of the highest quality in the kingdom — a result of the demand and the considerable paying power of the seventeen colleges and their inhabitants. Much of it came via the river on barges. The punts so beloved of students and unwieldy tourists today are the descendants of flat bottomed boats used to navigate the shallow marshlands of the region. Graham Swift's 1983 novel *Waterland* illuminates the part that water has played in the region. The university, which sent two members to Parliament, regulated the markets and issued licenses to trade with the university.

View of Cambridge. Drawn by R. Harraden, etched by Letitia Byrne 1809.

"Cambridge from the Ely Road". R. Ackermann's, *History of Cambridge*, 1815.

The Market Place shewing the Town Hall and Hobson's Conduit. T. D. Atkinson. *Cambridge described and illustrated.* Macmillan, 1897.

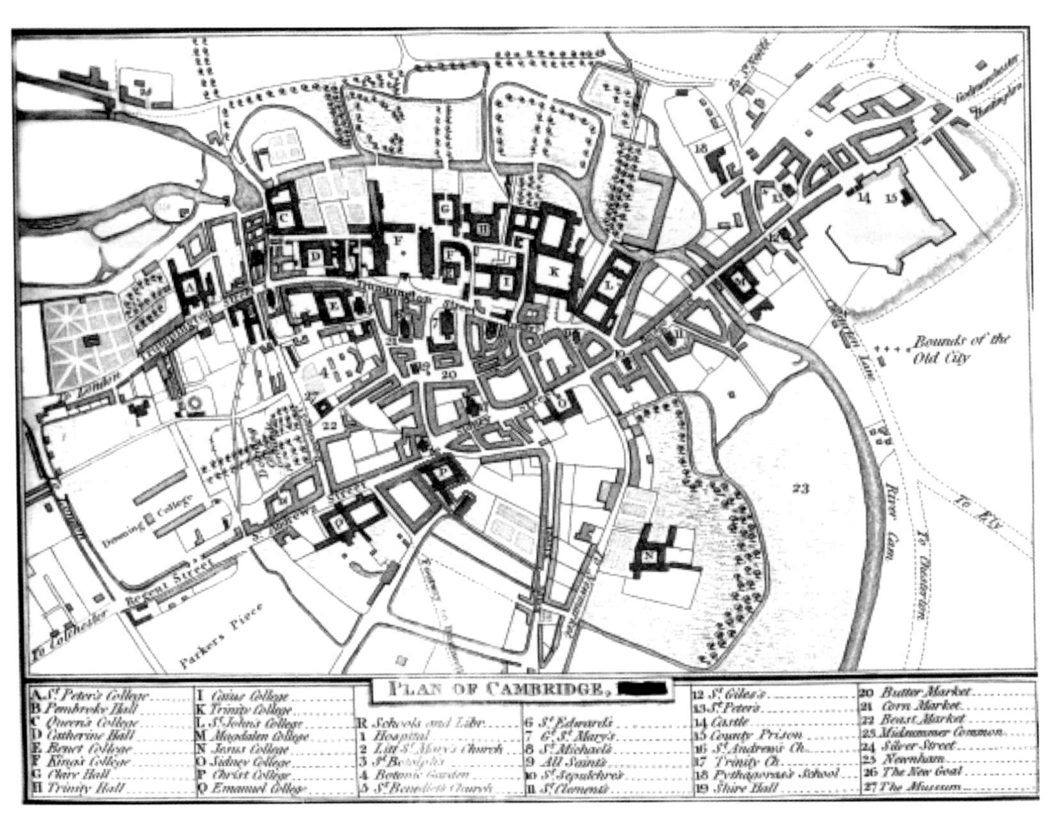

Map of Cambridge from *The Cambridge guide*, 1830. Christ's College has here been highlighted in red.

The colleges were like independent fiefdoms, each with its own head, usually called the Master, and a number of Fellows. There were about 400 college Fellows in Cambridge. Some of these served various roles in their college such as tutor, lecturer, steward, bursar or dean. Most of them were ordained clergymen. If a Fellow took a college living, that is vicar of one of the parishes controlled by that college, or married, he lost his Fellowship.

The students were ranked in three or four types. The first were Noblemen, the eldest sons of Dukes, Marquesses, Earls, Viscounts or Barons, of which there were so few that they were often not counted as a separate type of student. The next type, Fellow Commoners, were usually the younger sons of noblemen or otherwise extremely wealthy students who ate with the Fellows and sat with them in chapel. Hence, they were in common with them. The most numerous type of student was Pensioners; those able to pay all their own fees and expenses. The last group were Sizars; these depended on scholarships or other financial support and were often required to perform menial tasks in the college such as serving other students at meal times.

Members of the university, were required to wear academic dress (what is now often called 'cap and gown') in the streets and precincts of the university, though there were times of day and activities (such as riding) when this was excepted. As the *Cambridge guide* noted in 1830: "There are few objects that attract the stranger's notice more powerfully than the various academical dresses worn by the members of the University, and the appearance of so many men, of every age and rank, moving to and fro, arrayed in robes of various hues and shapes, is particularly striking; especially during full-term."[14]

Doctors of Divinity, Civil Law, Music and Medicine each had a distinctive gown, indeed gowns, and so did Bachelors of Arts (graduates). These could wear their gown with a normal hat rather than an academic cap or mortar board. The different ranks of students, Noblemen, Fellow Commoners, Pensioners and Sizars also wore different gowns which indicated their rank and in some cases their particular college. They were required to wear academical caps. Darwin wore a Pensioner's gown which was made of "black Princes' stuff, with square velvet collar, and facings, but without sleeves; and a square cap of black cloth, with silk tassel."[15] Prince's Stuff was yarn made of wool combined with silk used for making black clerical gowns, mourning attire and so forth.

[14] *The Cambridge guide*, 1830, p. 29.

[15] *The Cambridge guide*, 1830, p. 33.

A Pembroke Pensioner's gown (left) and a Christ's Pensioner's gown (right). *The costumes of the members of the University of Cambridge*. London: N. Whittock, c. 1840.

Amongst the University officers were four proctors, appointed yearly from amongst the Fellows. Some historians have claimed that the proctors wore non-distinctive gowns so that they could prowl the streets in disguise in order to enforce university regulations.[16] In fact, the office of proctor, like most other University offices, did not have a special gown. Proctors wore their usual college Fellows' or MA gowns.[17] Amongst their various duties, they most notoriously tracked down students carousing with young women. The proctors had the authority to enter any house in Cambridge where they suspected such types might be concealed; they also had the power to imprison any such young women in the 'spinning house', the University's own gaol. With so many well-to-do young men confined to all male colleges, it is hardly surprising that temptations were never far away. The name of the street "Maid's Causeway" retains its name from such times.

We have no evidence that Darwin ever sampled any of these offerings. But then little evidence is likely to have been left behind anyway. For example, no one dreamed that his cousin William Darwin Fox was up to no good. But his recently

[16] Desmond, Adrian and Moore, James. 1991. *Darwin*. London: Michael Joseph, p. 50.

[17] *The Cambridge guide*, 1830, p. 34.

published diaries, with comments about a girlfriend or servant girl written in Greek code, show that he was. His financial records reveal regular gifts of 1 shilling to a beggar, which was probably code for a prostitute.[18]

"Quite Unexpected." A proctor surprises an undergraduate with an illicit female visitor. Note the cap and gown on the floor in the foreground. Such illustrations were vital in restoring Darwin's rooms in 2009. *Gradus ad Cantabrigiam; or, the new university guide* (1824).

The academic year, then as now, was divided into three terms. Michaelmas: October–December; Lent: January–March; and Easter: April–July. An ordinary Bachelor of Arts (B.A.) degree required three and a third years (or ten terms) residence in Cambridge and the successful completion of two examinations, except for noblemen who simply 'proceeded' to their degree after the required residence. Students were required to spend each night in their college rooms or in approved and licensed lodgings if living outside college, as a means of accurately recording their required residence. An 'exeat' from the college Tutor could be acquired to allow for temporary leave.

Attendance at college lectures and readings was usually compulsory and could be enforced by fines and other punishments. Study in Cambridge was, however, encouraged through the incentives of awards or prizes rather than the fear of punishment. Christ's College, for example, offered annual cash and book prizes for the best composition in Latin verse and prose. Another set of prizes was offered "for the best Latin dissertation on some evidence for Christianity ... English Composition

[18] Larkum, Tony. 2009. *A natural calling: life, letters and diaries of Charles Darwin and William Darwin Fox.* Dordrecht and New York: Springer, p. 17.

on some Moral Precept of the Gospel" ... and "the most distinct and graveful reader in, and regular attendant at, Chapel."[19] University lectures were, in contrast, optional. Unlike Edinburgh, the focus in Cambridge was on reading and tutorials rather than lectures. The curriculum was divided into three general areas of study: natural philosophy, moral philosophy and classics. A contributor to the *Monthly Magazine* called it a "very liberal system of education ... exquisitely adapted to rouse genius into energy and sluggishness into action".[20] All this was to prepare young men for the church, for the Bar or for politics. Women could not attend Cambridge until 1870.[21]

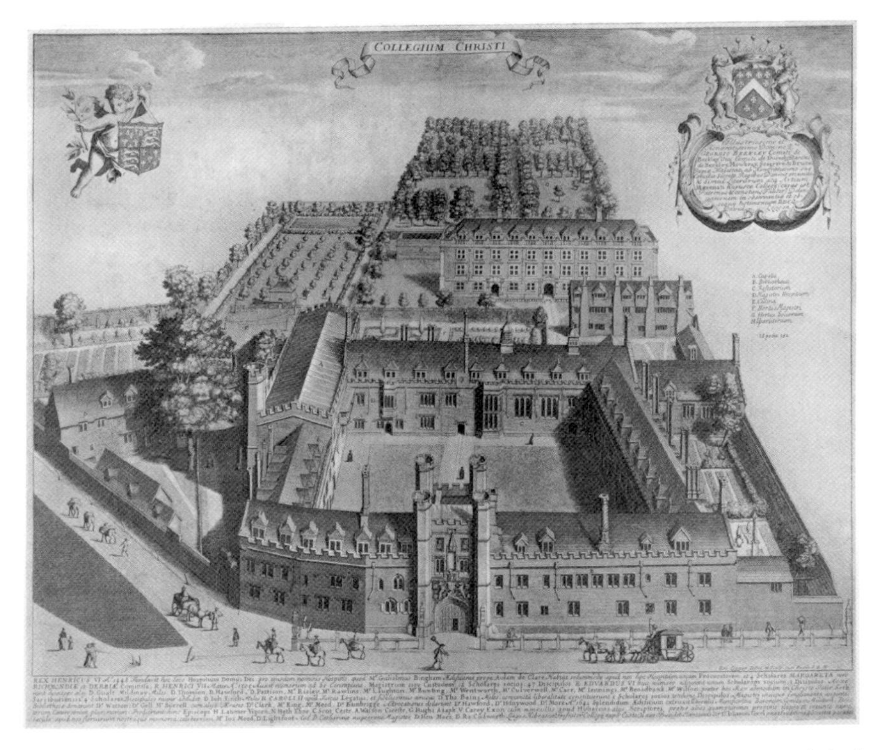

Bird's eye view of Christ's College from D. Loggan's *Cantabrigia Illustrata*, 1690. Little had changed by Darwin's time.

[19] *Cambridge University Calendar* 1829, pp. 286–287.

[20] Alpha Beta. 1803. An account of Cambridge. *Monthly Magazine and British Register, 15 (February, March)*, p. 27.

[21] For more details on the University curriculum and the reforms it experienced in the first half of the nineteenth century see Garland, M.M. 1980. *Cambridge before Darwin: the ideal of a liberal education,* 1800–1860. Cambridge: University Press.

Christ's College

Christ's College is one of the smaller colleges in Cambridge and faces St. Andrew's Street, directly across from the Church of St Andrew the Great. Christ's, replacing the earlier God's House (1439), was re-founded in 1505 by Lady Margaret Beaufort (1443–1509), the mother of Henry VII. Lady Margaret endowed her new institution handsomely with lands, estates, ecclesiastical livings and valuable gold and silver objects.

The original buildings in First Court, consisted of the Great Gate, the Chapel, the Master's Lodge, the Hall, the kitchen, the Library and accommodation. These date to the mid-fifteenth and early sixteenth centuries. Originally, the College resembled somewhat the current appearance of St John's College in that it was red brick and clunch. Clunch is a white chalk stone of East Anglia. In 1758–1769, the exterior of Christ's first court was faced with Ketton oolite stone which remains the visible exterior surface today. Only the outer south side wall, which then faced what was known as bath court, was left uncovered. By passing through the passage to the Undergraduate library, it is possible to see the original clunch and red brick.

Newer buildings in Second Court provided additional accommodation including the Fellows building begun in 1640. Farther back there were the College gardens which contained shady walks, decorated with busts and alcoves,

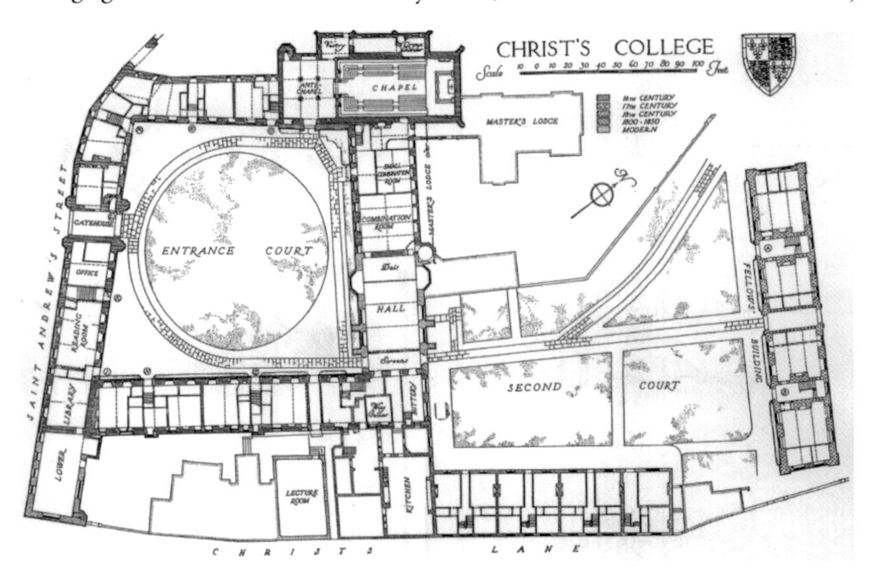

Plan of Christ's College from *An inventory of the historical monuments in the city of Cambridge* (1959).

"Milton's mulberry-tree" and "a bath, with an elegant summer-house" in Christ's College garden. T. Wright, *Memorials of Cambridge* (1845).

a bowling-green, a summer house and a bathing pool. The guidebooks of the 1820s and 1830s recommended the ancient mulberry tree which was sometimes said to have been planted by the great poet and alumnus of the College, John Milton (1608–1674). Another story is that the poet sat under the tree writing some of his poetry. More creditably it is believed to have been planted in the year of Milton's birth, 1608. It can still be seen today supported by many props. It continues to furnish the Fellows with a large crop of mulberries each year.

In the 1820s Christ's had fifteen Fellows, most of whom were required to take Holy Orders within twelve months of attaining the proper age. Amongst the students and resident M.A.s and B.A.s there were (numbers fluctuated each year) about four Noblemen, sixteen Fellow Commoners, sixty-five Pensioners and eight Sizars. The names of the Master, Fellows and students were listed each year in the *University Calendar*.[22] On average Christ's admitted thirty new undergraduates each year. The entire university admitted about 400 each year.

In principle, the College was still governed by the Elizabethan statutes (these were not changed until 1860) but in reality Cambridge colleges had long since gradually changed their day-to-day practices to suit changing times. Christ's subsisted on the rents of about eighty properties from the endowment of the Foundress and, in turn, bestowed the clerical livings of seventeen parishes. The livings were supported by the second tax that people paid to the church — tithes and rents.

Why Christ's College?

The reason Darwin came to Christ's College has long been debated. His grandfather, the poet Erasmus Darwin, went to St. John's College. Shrewsbury School also had connections to St. John's. But the reasons that brought Darwin to Christ's are plain enough. In 1821, his cousin, Hensleigh Wedgwood (1803–1891), later a philologist and barrister, also went to St. John's. At the time it had a reputation for strict discipline. Wedgwood migrated to Christ's after only a single term. He

[22] The usual invaluable resources for identifying individuals at Cambridge and Christ's respectively are: Venn, J. A. 1940–1954. *Alumni Cantabrigienses*. Part II, 1752–1900. Cambridge: University Press and Peile, J. 1913. *Bibliographical Register of Christ's College, 1505–1905*. vol. II 1666–1905. However these tend to list only those who graduated. The contemporary *University Calendar* is an excellent resource for identifying who was at each College and to see which offices were currently held by particular Fellows. Those issues for Darwin's years at Christ's are reproduced on *The Complete Work of Charles Darwin Online* (http://darwin-online.org.uk).

took his B.A. in 1824 and was elected a Finch and Baines Fellow of Christ's in February 1829, a position he held until October 1830. This was one of the few Fellowships that did not require taking Holy Orders. Hence it is not surprising that his cousin, Darwin's brother Erasmus, joined the not-too-demanding Christ's College on 9 February 1822. He received his M.B. in 1828. According to Darwin's son Francis: "The written part of the examination for the M.B. consisted in being left alone with a paper of questions in the Regius Professor's library while that official went to see a patient at some distance from Cambridge."[23]

Darwin's second cousin, William Darwin Fox (1813–1881), later a clergyman naturalist, came up in 1824.[24] It was therefore perfectly natural that Darwin would follow his cousins and brother to Christ's. And equally that he join a college amenable to wealthy young men devoted to hunting and shooting.

In Darwin's day Christ's was a quiet and relaxed institution, neither academically rigorous nor religiously strict. Darwin's son, Francis, later recorded:

> The impression of a contemporary of my father's is that Christ's in their day was a pleasant, fairly quiet college, with some tendency towards "horsiness"; many of the men made a custom of going to Newmarket during the [horse] races, though betting was not a regular practice. In this they were by no means discouraged by the Senior Tutor, Mr. Shaw, who was himself generally to be seen on the Heath on these occasions. There was a somewhat high proportion of Fellow-Commoners, — eight or nine, to sixty or seventy Pensioners, and this would indicate that it was not an unpleasant college for men with money to spend and with no great love of strict discipline."[25]

Former Master of Christ's, Sir Arthur Shipley, recalled in 1924: "It is in the recollection of those now living that one of the College 'gyps' [servants] used to recount how when young he had seen a number of students in scarlet coats ride round the first court."[26]

---

[23] [Darwin, Francis (ed.)]. 1909. Some letters from Charles Darwin to Alfred Russel Wallace. Darwin Centenary Number. *Christ's College Magazine* 23 (Easter Term): 229.

[24] On Fox see Larkum, Anthony. 2009. *A natural calling: life, letters and diaries of Charles Darwin and William Darwin Fox*. Springer. This important work also includes new annotations to the letters from Darwin to Fox and extensive transcripts from Fox's 1824–1826 diary. The entire diary is reproduced in facsimile form on *Darwin Online*: http://darwin-online.org.uk/content/frameset?viewtype=image&itemID=CUL-DAR250.5&pageseq=1.

[25] Darwin, F. 1887. *The life and letters of Charles Darwin*, vol. 1: 65.

[26] Shipley, A.E. [1924]. *Cambridge cameos*, p. 126.

College Administration of Students

The Tutor recorded the entry of students, received their tradesmens' bills, kept their accounts and generally looked after them. The Tutor was paid through his banker. Darwin was admitted as a member of the College, presumably by post, under Tutor Joseph Shaw (1786–1859). Shaw, politically one of the liberal-minded reformers in the College, had been admitted as a Sizar at Christ's in 1803, gaining his B.A. in 1807 and M.A. in 1810 and was a Fellow of the College between 1807 and 1849. A College historian wrote that Shaw was:

> obviously an arch-snob, [who] undoubtedly helped things along by laying himself out to make Christ's particularly attractive to ex-officers of the richer classes, fellow-commoners and noblemen, together with well-to-do young men of any sort and in general those of sporting rather than of intellectual tastes … In the whole time [1814–1828] only 49 of those entered under him took Honours, of whom 12 belong to the three years when he had ceased to be Tutor. Only three sat for the Classical Tripos …[27]

Darwin's name was entered in the Admissions Books at Christ's College on 15 October 1827. There are three surviving Admissions Books which record Darwin's admission. The specific purpose of each book, apart from the main Admissions Book, has not been ascertained. One might have been kept by the Bursar, for example, and another by the Tutor.

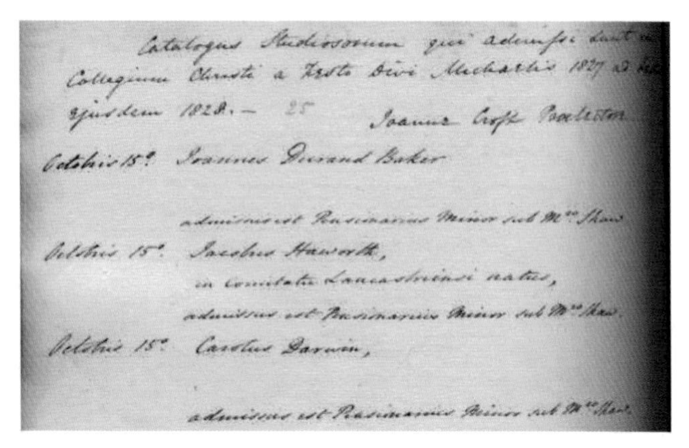

Catalogus Studiosorum qui admissi sunt

Collegium Christi a Festo Divi Michaelis 1827

ejusdem 1828. — 25 Joanne Croft Praelector

Octobris 15º Joannes Durand Baker

admissus est Pensionarius Minor sub Mro Shaw

Octobris 15º Jacobus Haworth,

in Comitatu Lancastriensi natus,

admissus est Pensionarius Minor sub Mro Shaw.

Octobris 15º Carolus Darwin,

admissus est Pensionarius Minor sub Mro Shaw.

'Admissions to Christ College' (T.1.2)

[27] Steel, Anthony. 1949. *The custom of the room or early wine-books of Christ's College*. Cambridge, pp. 40–41, see also Peile, J. (ed.) 1913. Biographical register of Christ's College 1505–1905 and of the earlier foundation, God's House 1448–1505. 2 vols. Cambridge: University Press.

Catalogus Studiosorum qui admissi sunt in Collegium Christi a Festo Divi Michaelis 1827 ad Festum eiusdem 1828. –
Octobris 15. Carolus Darwin,
admissus est Pensionarius Minor sub M^{ro} Shaw.
(Translation: List of students who are admitted to Christ College from the Feast of Saint Michael [29 September] 1827 to the same Feast in 1828. –
October 15. Charles Darwin,
admitted as Minor Pensioner under Master Shaw.)

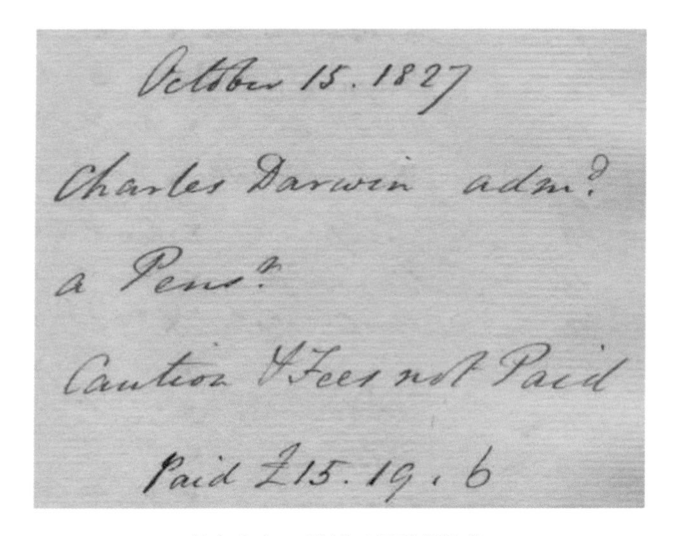
October 15. 1827
Charles Darwin adm^{d}
a Pens^{r}
Caution & Fees not Paid
Paid £15. 19. 6

'Admissions 1818–1828' (T.1.4)

October 15. 1827
Charles Darwin adm^{d}. a Pens^{r}.
Caution & Fees not Paid
Paid £15.19.6

"Paid £15.19.6" was added later, presumably when Darwin arrived in Cambridge or when payment was received by cheque. The Caution money (£15) was a deposit to the Tutor against good behaviour and unpaid bills and so forth, returned at graduation. The fees made up the remaining 19 shillings and sixpence.

This Admissions Book, described in the catalogue of archives as 'Admissions/ Terms Books', records Darwin's residence in Cambridge and the date of his B.A. degree and the date of his death in 1882.

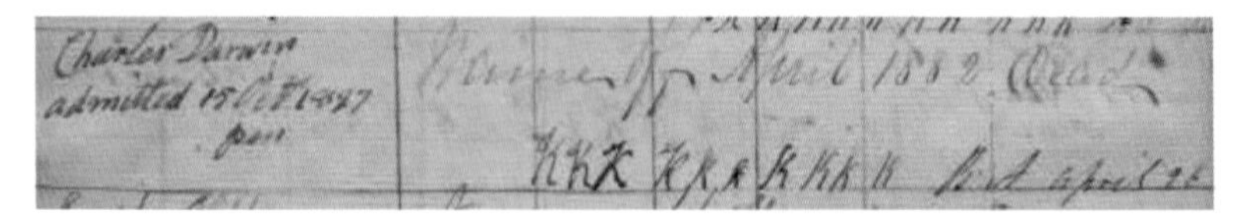

'Admissions 1815–1852' (T.3.1) The "k's" stand
for terms 'kept' or resided in Cambridge.

There are also six College record books, recently re-discovered by the Honorary Keeper of the Archives and Fellow Commoner,[28] Geoffrey Thorndike Martin, which reveal previously unknown details of Darwin's time at Christ's. For example, we learn that he occupied one of the most expensive range of rooms for an undergraduate, that he paid extra for vegetables with dinner and many other financial and household details. There is even a fragment of what may have been a medical certificate of illness or sick note for Darwin. Because students like Darwin often paid local tradesmen by account, their individual bills were reported to the College. Hence the Books reveal Darwin's accounts not just his College but for the barber, grocer, tailor, laundress, chimney sweep and much more. The six record book details for Darwin are given in the Appendix.

'1822–1829 Tutors' Accounts' (T.11.26) and '1830–1835 Students Bills' (T.11.27) record the payment of Darwin's Tutor's bills during his time at Christ's:

£ s d	
55.16.06	J.S. Apr. 22 1828
106.05.0	J.S. July 1. 1828
15.18.06	J.S. Nov. 21. 1828
39.15.03	J.G. March 5. 1829
26.15.03	J.G. May 29. 1829
58.05.07	J.G. July 1829
70.18.07	J.G. March 8. 1830
49.10.00	E.J.S. June 4. 1830
[53.09.11]	Midsummer 1830
122.05.09	May 1831
33.12.6½	[June 1831][29]
£636.0.9½ in total	

'J.S.' = Joseph Shaw and 'J.G.' = John Graham. E.J.S = Edward John Ash.

[28] The term Fellow Commoner has been revived for a type of honorary Fellow. It should not be confused with the rank of student in Darwin's day.

[29] This bill was disputed see Darwin to Caroline Darwin [31?] October [1831] in *Correspondence*, vol. 1, p. 177.

Name	Residence													Total	Residence		
Clarke jun.	in 14 March out 27 March in 28 April out 2 May in [illegible] in 15 April out 29 May	7·11¼	5 9¼				2 6¾	7·9¾	4·6¼	8·8¾	7·6½	7·3¼	5·9¾	0·5½	2 18·5	in 15 March out 22 March in 23 April out 3 May	55 4·0·5½	1·5·9½
Holroyd						2·10¾	6 0¼	7·7¼	6·9	7·7¼	5 9¾	15·8¼			2·12·0¾	in 16 April out 31 May	3 5½ ·6½	1·3·10
E. F. Smith	in 14 March	6·9¼	10·2¼	6·6¾	6·3¼	3·5¾	6·7¼	9·8¼	8·6¼	9·5¾	7 8	11·1½			4·6·4¾	in 15 March	7¾ 5·6·2½	1·13·11
Jackson	in 14 March	5·11½	5 8¾	7·0¼	5 5¾	8·1¾	7·6¼	11·3¾	7·5¾	7·11¾	7·3¾	12·10¾	0·7		4·7·4½	in 15 March	7½ 5·6·2½	1·13·11
Rickards	in 14 March out 26 March in 19 April ~~in 14 March~~	6·0¾	5 4¾				5·10¼	7·10¼	5 9¾	7·0¾	6 0¾	7·0¼	0·4½		2·11·[illegible]	in 15 March out 26 March in 19 April	5¼ 4·0·4	1·5·8
Dawson																		
Graham	in 19 April out 29 May						4·0¾	6·0¼	5·10¾	2·5¾	5·0½	5 5¼			1·8·11¼	in 19 April	[illegible] 3·0·3	0·12·3
Baker	in 14 March out 19 March in 21 April	3 5¼					4 5¼	6·10¾	6 1¼	5 4¾	6 1¼	5 4¾	2·3		2·0·1¾	in 15 March out 19 March in 23 April	4½ 3·0·4½	0·19·3
Haworth																		
Darwin	in 14 March out 25 March in [illegible] March	9·1¾	5·9¼	7·8¼	7·9	9 8¼	10·5¾	14·1¾	12·6¼	17·7¾	[illegible]·1¾	13·1			5·17·10¾	in 15 March ~~[illegible]~~	7¼ 5·6·9½	1·13·11
Cherriere	in 14 March out 5 April	11·5¾	12·1	13·2	5 9¼	–									2·2·6	in 15 March [illegible]	14 1·1·1	0·6·5

This page from '1827–1831 Students Bills' (T.11.25) is the only known document that records the date of Darwin's arrival in Cambridge. It also records his weekly commons or buttery account and extra vegetables in the final column.

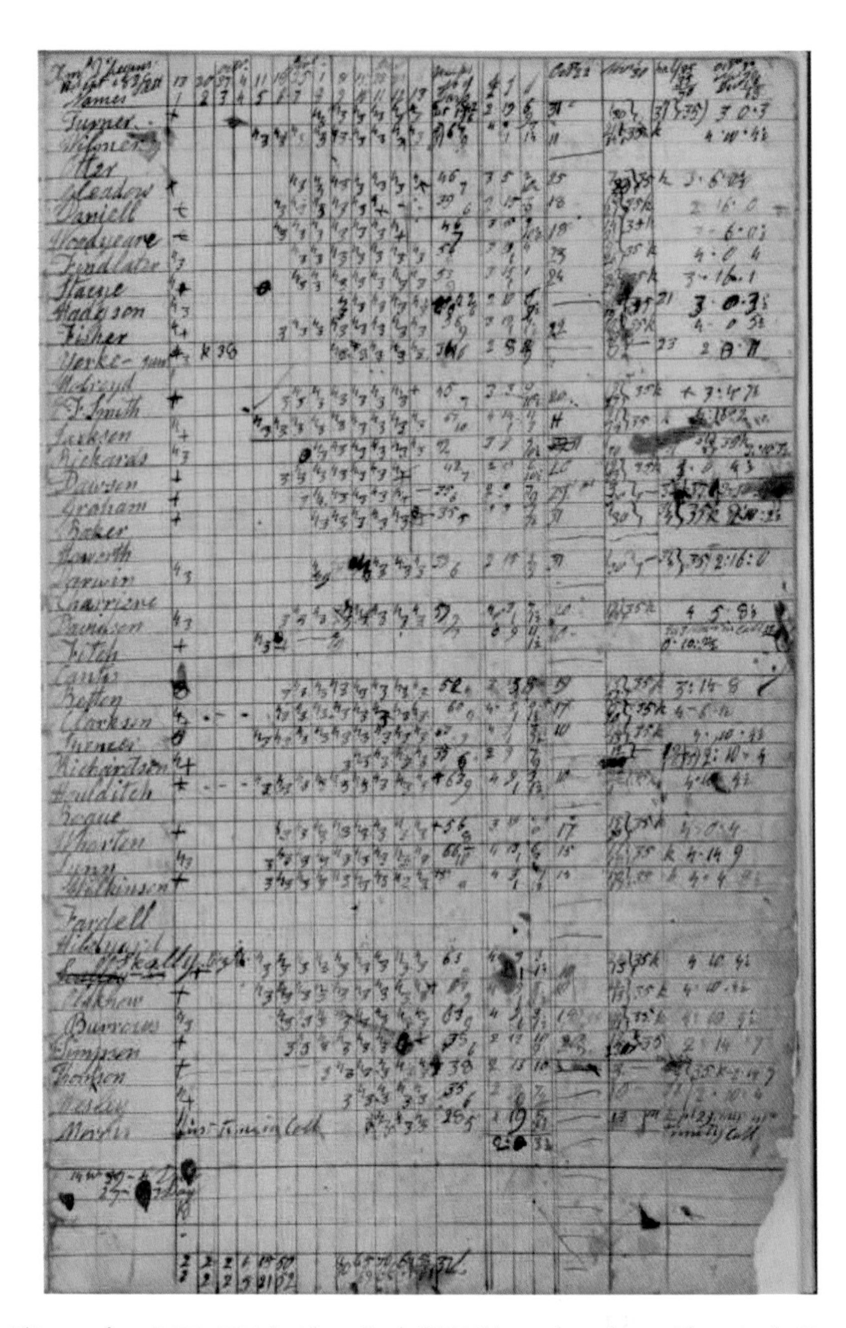

This page from '1828–1829 Residents Book (T.17.A)' records student residence in the Xmas quarter, September–December 1828.

The itemized quarterly bills added together give the following totals:

15.3.0	Barber
12.1.6	Bedmaker
1.7.0	Bookseller
0.8.6	Carpenter
57.11.0	Chamber
0.4.0	Chimney sweep
28.0.1	Coals
78.0.0	Cook
5.11.0	Eve Joiner
4.12.6	Glazier
39.17.0	Grocer
12.19.6	Laundress
20.11.0	Library
8.5.0	Painter
16.19.9½	Porter
1.13.6	Scullion
0.1.6	Seamstress
5.3.6	Shoeblacker
15.12.6	Shoemaker
0.8.6	Smith
125.16.7	Steward
35.11.6	Taylor
37.10.0	Tuition
40.5.6	Woollen draper [one entry in 1828]

The Record Books at Christ's College are nevertheless not the complete picture of Darwin's finances at Cambridge. Additional expenses we know of include a horse he kept from 1830 although no records have been found for stables. Darwin's father gave him £200 to settle debts in Cambridge in 1831, any debts paid with this sum do not appear to be recorded at Christ's College. Similarly Darwin's gyp (college servant similar to a valet) does not appear on the bills, which indicates that the gyp was paid by the College and not the student.

Darwin's father probably paid the bills by cheque, and also supplied Darwin with an allowance of £300 according to Janet Browne.[30]

Darwin's Terms and Degree

It is very often stated that Darwin studied theology or divinity at Christ's. This is not correct. Although the reason Darwin came up to Cambridge was ultimately to become a clergyman, he was enrolled as a candidate for an ordinary Bachelor of Arts degree, or B.A. After which he could have taken divinity training prior to taking Holy Orders. Darwin never undertook the divinity training. There were as yet no science degrees. To earn a B.A. at Cambridge it was necessary to reside ten terms in Cambridge (or within a mile of Great St Mary's, the University church) and pass two university examinations, the 'Previous Examination' in the second year and the B.A. Examination in January of the final year. Darwin's terms, taken from the *Cambridge University Calendar*, are given below beside the dates Darwin was in residence according to the College record books:

1827

Michaelmas: 10 October–16 December [name added to Admissions Books 15 October 1827]

1828

Lent: 13 January–29 March ["in 26 Jan" T.11.25 "in 14 March–out 15 March–in 17 March" T.11.25]

Easter: 16 April–4 July 1828 [no entries]

Michaelmas: 10 October–16 December ["in 31 Oct", in commons: "1 Nov" T.11.25]

1829

Lent: 13 January–10 April [in "Feb 24", in commons: 25 Feb T.11.25]

Easter: 29 April–10 July ["out June 8" T.11.25]

Michaelmas: 10 October–16 December [in "Oct 12" T.11.25]

[30] Browne 1995, p. 90. I have been unable to locate a source for this sum, although she may have referred to Erasmus Darwin. Compare a contemporary student bill from Trinity College, as printed in Wright, John M.F. 1827. *Alma Mater, or, seven years at the University of Cambridge*. London: Black, Young, and Young, p. 112.

1830
Lent: 13 January–2 April ["in Jan 1" T.11.25]
Easter: 21 April–9 July ["out June 3" T.11.25]
Michaelmas: 10 October–16 December ["in Oct 7" T.11.25]

1831
Lent: 13 January–25 March
Easter: 13 April–8 July ["B A April 25th" [sic] T.11.25
"out June 16" T.11.25]

Darwin arrived in Cambridge, as the College records reveal for the first time, on Saturday 26 January 1828. He was eighteen years old. As the academic year began the previous October, all College rooms were already full. He therefore took lodgings above the premises of William Bacon, tobacconist, in Sidney Street, less than a minute's walk from the College Gate. All student lodging houses had to be licensed by the University and landlords were required to record the time when their lodgers returned in the evenings, thus maintaining the sort of records kept by the porters at college gates.

Modern histories have given no explanation for why Darwin lodged over Bacon's. The 1911 *College Magazine*, however, reveals interesting details that seem to have been overlooked.

> the exact whereabouts of [Darwin's lodgings] has been involved in considerable doubt. When, later in life, Darwin revisited Cambridge he one day pointed out to his son (now Sir George Darwin) the house in Sidney Street, in which, he said, he had once had rooms. He added that at that time Bacon the tobacconist occupied the shop below. Unfortunately Sir George Darwin was unable in after years to identify the house, and it is only recently that the efforts that have been made to ascertain its exact position have been crowned with success. It has now been proved that it was the house which stood on the site now covered by No. 63, Sidney Street, and occupied by Messrs Rutter and Son.
>
> The person to whose wide local knowledge the admirers of Darwin are indebted for the settlement of this interesting point is Mr Thomas Hunnybun … Mr Hunnybun was born in the year 1830 in the house on the opposite side of the street to the present No. 63. … Mr Hunnybun remembers that the house now occupied by Messrs Rutter and Son was once tenanted by Bacon the tobacconist, and that the latter used to let lodgings to University undergraduates, generally to Christ's men. Mr Hunnybun says that he well remem-

> bers his father going on several occasions to call on the Master of Christ's College to complain of the behaviour of undergraduates at Bacon's; and this fact impressed the recollection of the house upon his mind. The cause of complaint was, apparently, that the sporting young gentlemen over Bacon's were in the habit of leaning out of the windows and with tandem whips flicking the passers by. This seems to have been a favourite amusement at the time, for it is said that sixty years ago it was quite dangerous to walk down Rose Crescent on a Sunday morning, and one was unlikely to get through without having one's hat literally whipped off.[31]

The houses along Sidney Street, including Darwin's lodgings, were demolished in the 1930s. Today, two plaques on the site, occupied by a branch of Boots the Chemists, recall its approximate location. No biography of Darwin has ever shown the appearance of the original building. *Spalding's Cambridge Street Directory* for 1887 lists the properties as follows, working south from Market Street: Holy Trinity Church, Cambridge Savings Bank [No number given], 63 Rutter, Arthur, Philo Chambers.

Formal admission to the University followed Matriculation. The word derives from the requirement that undergraduates enter their names on the 'matricula' or role. The correct date of Darwin's matriculation has not been previously published. Darwin biographers Desmond and Moore, for example, state that it was "one morning in January 1828".[32] The first University matriculation ceremony after Darwin's arrival in Cambridge was Ash Wednesday, 20 February 1828. On that day, five men from Christ's matriculated. Around one o'clock Freshmen, grouped by college, gathered in the Senate House and signed their names in the Registrary's book under a heading declaring they were bona fide members of the Church of England as by law established. Another young man who signed that day was "Alfred Tennyson" of Trinity College later to become the Poet Laureate Alfred Lord Tennyson. Darwin signed as "Charles Robert Darwin". They took the Latin oaths of allegiance and supremacy before the Senior Proctor, that year Woodwardian Professor of geology Adam Sedgwick (1785–1873). Thus began Darwin's lifelong association with the University of Cambridge.

---

[31] Anon. 1909. Darwin's lodgings. *Christ's College Magazine* (Easter Term), pp. 248–249.

[32] Desmond, Adrian and Moore, James. 1991. *Darwin*. London: Michael Joseph, p. 52.

Darwin's lodgings (indicated by arrow) on Sidney Street. *The Graphic*, 8 October 1887, p. 404.

A photograph of Sidney Street from the Royal Agricultural Society's Royal Show programme for 1922. Both images are courtesy of Christopher Jakes, Principal Librarian, Local Studies, Cambridgeshire Libraries.

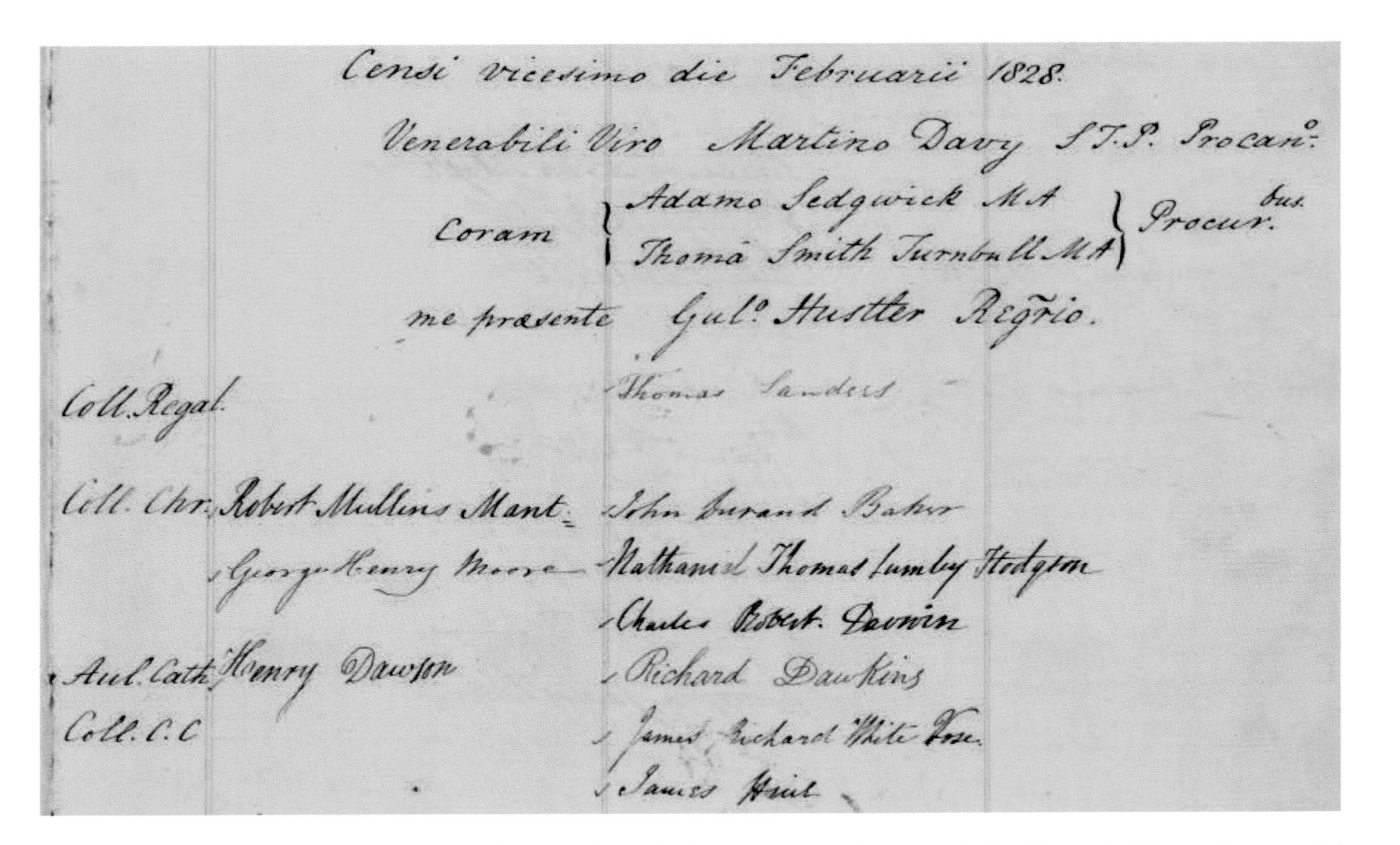

Censi vicesimo die Februarii 1828.

Venerabili Viro Martino Davy S.T.P. Procan^o:

Coram { Adamo Sedgwick MA / Thoma Smith Turnbull MA } Procur:^bus

me praesente Gul^o Hustler Regrio.

Coll. Regal. Thomas Sanders

Coll. Chr. Robert Mullins Mant — John Durand Baker

George Henry Moore — Nathaniel Thomas Lumby Hodgson

Charles Robert Darwin

Aul. Cath. Henry Dawson — Richard Dawkins

Coll. C.C. James Richard White Vose

James Hail

Darwin's signature in the 'Books of subscriptions for degrees', Cambridge University Archives, Cambridge University Library. Courtesy of the Syndics of Cambridge University Library. (Matric. 11.) Amusingly, the student who signed after Darwin was named Richard Dawkins.

The interior of Senate House. Drawn by R. Harraden and engraved by J. Skelton, 1810.

FIRST YEAR AT CAMBRIDGE

Settled in his lodgings overlooking what was then a much narrower Sidney Street, Darwin began to make friends and renew old acquaintances. His brother Erasmus returned on 8 February 1828 (T.11.25). Perhaps lodging over the tobacconist's shop led to some teasing. One friend, Albert Way (1805–1874) of Trinity College, drew a mock coat of arms for Darwin in April 1828. This comic coat of arms depicts crossed tobacco pipes, meerschaum pipes, cigars, a wine barrel and beer tankards, evidently drinking and smoking were Darwin's trademarks!

Car[olo] Darwin
amico opt[imo] amicus fumosior
Albertus Way e C[ollegio] S[anctae] Trin[itatis]
D[at] D[icat] D[edicat]
5 Ap[ril] 1828

Translation: Albert Way of Trinity College, a more smoke-filled friend, gives, devotes and dedicates [this] to his best friend Charles Darwin.

David Butterfield, classicist and Fellow of Christ's College, provided the following explanation:

> 'fumosior' implies that both Darwin and Way are lovers of tobacco, but that Way is the smokier of the two. 'Yalo Baccoque repleti', means '[we are] filled with Yalus and Baccus'. Bacc(h)us is the god of wine and is the figure standing on the right; his name

Albert Way's comic coat of arms for Darwin. From (DAR 204:30). Courtesy of the Syndics of Cambridge University Library.

is often used metonymically for wine. 'Yalus', therefore, is presumably represented by the figure on the left. The figure looks like an American Indian. In Lakota (the language spoken by the Sioux) yalu-icu means to suck smoke through a pipe.

The joke probably stems from the fact that Darwin was apparently not much of a drinker at Cambridge. In later life, he barely drank wine with meals and his university friends thought he exaggerated in his autobiographical recollections of drinking too much a few times. Darwin did take snuff, which he continued to do for the rest of his life.

At any rate, Darwin also cultivated more sedate interests and opportunities available at Cambridge. Never interested in team sports, he did not join the cricket matches on the wide green expanse of Parkers piece or the new sport of rowing on the Cam.

> I also got into a musical set, I believe by means of my warm-hearted friend Herbert, who took a high wrangler's degree. From associating with these men and hearing them play, I acquired a strong taste for music, and used very often to time my walks so as to hear on week days the anthem in King's College Chapel. This gave me intense pleasure, so that my backbone would sometimes shiver. I am sure that there was no affectation or mere imitation in this taste, for I used generally to go by myself to King's College, and I sometimes hired the chorister boys to sing in my rooms. Nevertheless I am so utterly destitute of an ear, that I cannot perceive a discord, or keep time and hum a tune correctly; and it is a mystery how I could possibly have derived pleasure from music.
>
> My musical friends soon perceived my state, and sometimes amused themselves by making me pass an examination, which consisted in ascertaining how many tunes I could recognise, when they were played rather more quickly or slowly than usual. 'God save the King' when thus played was a sore puzzle. There was another man with almost as bad an ear as I had, and strange to say he played a little on the flute. Once I had the triumph of beating him in one of our musical examinations.[33]

John Maurice Herbert (1808–1882), later recalled that he took Darwin to King's College Chapel. "[Darwin] had great enjoyment in, & a keen relish for fine concerted music — both instrumental & choral, & we frequently went to King's College Chapel to hear the anthem in the Afternoon service…He was very fond of riding, & he sometimes hunted, but he never boated, or played cricket."[34]

---

[33] *Autobiography*, pp. 61–62.

[34] Herbert, J. M. 1882. [Recollections of Darwin, 12 June.] (DAR 112.A60-A61), transcribed by Kees Rookmaaker. *The Complete Work of Charles Darwin Online* (http://darwin-online.org.uk/).

"South side of King's College Chapel." R. Ackermann's, *History of Cambridge*, 1815.

Interior of King's College Chapel. R. Ackermann's, *History of Cambridge*, 1815.

Darwin recalled how a former Shrewsbury school friend, Charles Thomas Whitley (1808–1895), then at St John's, "inoculated me with a taste for pictures and good engravings, of which I bought some. I frequently went to the Fitzwilliam Gallery, and my taste must have been fairly good, for I certainly admired the best pictures, which I discussed with the old curator."[35] The Fitzwilliam Museum was then in Free School Lane, in what is now the Whipple Museum of the History of Science. The elegant edifice of the current Fitzwilliam Museum building was not begun until 1837. According to Herbert: "[Darwin] had a great liking for first-class line engravings — especially for those of Raphael Morghen & Müller; & he spent hours in the Fitzwilliam Museum in looking over the prints in that collection."[36]

At least one of the engravings Darwin had in his rooms still survives, and he seems to have displayed it in his study throughout his life. It is an engraving of Leonardo da Vinci by Raphael Morghen (1758–1833). It still hangs today in Darwin's restored Old Study at Down House. On the back is inscribed: "Gift of Leonard Darwin, 1929" and "This belonged to Father". So at least one example of Darwin's youthful interest in art still looked over his shoulder as he penned the *Origin of Species* in 1859.

Leonardo da Vinci, portrait with cap and fur collar. Raphael Morghen, c. 1817.

---

[35] *Autobiography*, p. 61. The curator was William Key Ridgway, d. 1852.

[36] Herbert, J. M. 1882. [Recollections of Darwin, 12 June.] (DAR 112.A60-A61), transcribed by Kees Rookmaaker. *The Complete Work of Charles Darwin Online* (http://darwin-online.org.uk/).

Engraving of Raphael Sanzio's Madonna della Sedia by Auguste Gaspard Louis Boucher Desnoyers. A copy of this engraving was owned by Erasmus Darwin. Reproduced courtesy of the Fitzwilliam Museum, Cambridge, acc. no. 33.A.1–107.

Engraving of Raphael Sanzio's Apollo on Parnassus, a mural in the Stanza della Signatura, in the Vatican by Marcantonio Raimondi. Possibly one of the engravings Darwin displayed in his rooms at Christ's. (See H. E. Litchfield's recollection in DAR 262.23: 9–10) Reproduced courtesy of the Fitzwilliam Museum, Cambridge, acc. no. P.5323-R.

Just as at Edinburgh, Erasmus soon moved on leaving his younger brother to fend for himself. Erasmus earned his M.B. degree and left for a grand tour of the continent. His bill was paid on 2 April 1828. Most of Darwin's friends were from other colleges, as he recalled in later life "I do not think I knew even to bow to 15 men in college & was intimate with only 2 or 3 men. — Most of my friends belonged to Trinity & St. Johns & Emanuel [sic]."[37] Perhaps Darwin's closest friend at Cambridge was his second cousin, William Darwin Fox, who was also reading at Christ's (admitted 26 January 1824) for an ordinary Bachelor of Arts degree with the aim of becoming a clergyman. Fox, again like Darwin, enjoyed riding, shooting and natural history. Fox was particularly fond of birds and insects. He kept a series of diaries which record in remarkable detail the life of a Christ's undergraduate of the time. Darwin's letters to Fox are now preserved in the College's Old Library. They provide the primary source for our knowledge of Darwin's undergraduate life. After morning service in the College chapel, Darwin often joined Fox for breakfast in his rooms just through the archway leading to Second Court in what is today E staircase. In after years, Darwin fondly remembered "our ancient snug breakfasts at Cambridge". "What pleasant hours, those were when I used to come & drink coffee with you daily!"[38]

[37] *Correspondence*, vol. 7: 38.

[38] *Correspondence*, vol. 1: 223; *Correspondence*, vol. 7: 196.

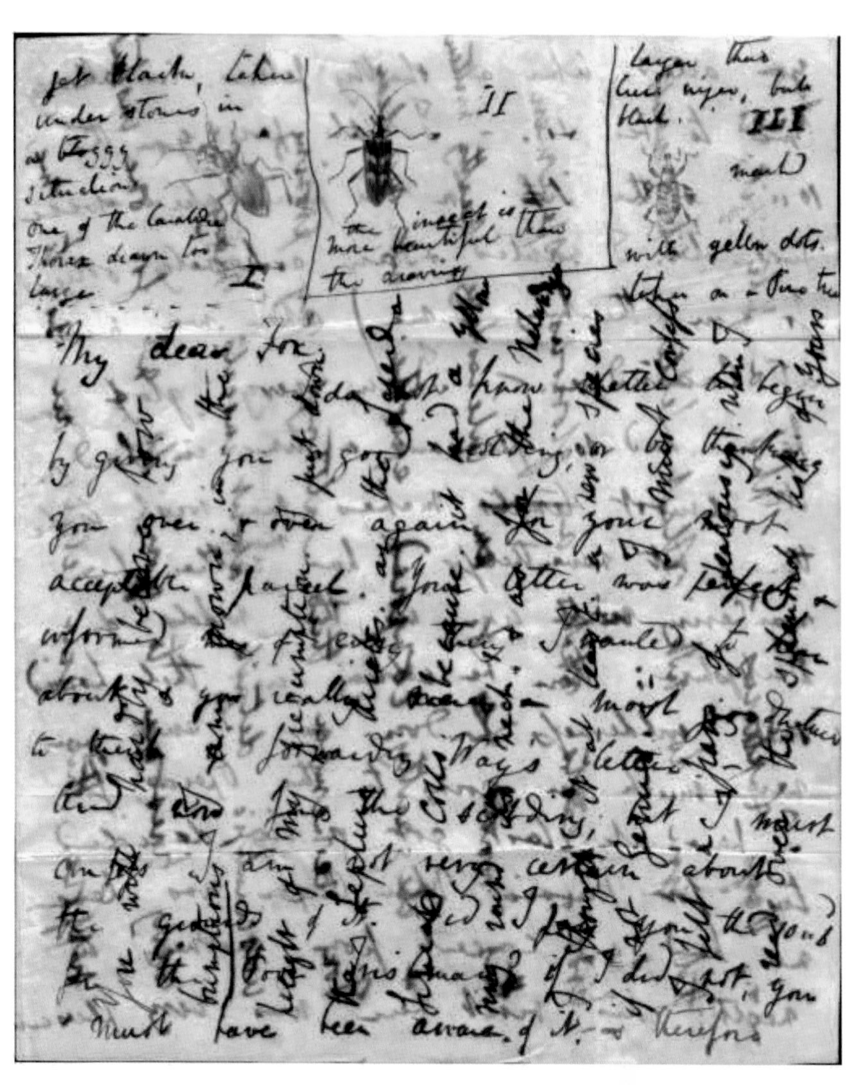

jet black, taken under stones in a Boggy situation
One of the Cambridge Thorax drawn too large
I

II
the insect is more beautiful than the drawing

larger than wings, but black. III
mark'd with yellow dots.

My dear Fox

Drawing of beetles by one of Darwin's sisters, in a letter by Darwin to W. D. Fox [30 June 1828]. Christ's College Library, Fox 1. See *Correspondence*, vol. 1: 58.

Christ's College New Buildings
CAMBRIDGE University ALMANACK 1832.

The Second Court of Christ's. Fox's rooms were on the first floor reached via the central doorway in the building on the left.

Fox inspired Darwin to embrace the scientific hobby of collecting insects, particularly beetles. Darwin was soon scouring the surrounding fields and fens for rare species.

> no pursuit at Cambridge was followed with nearly so much eagerness or gave me so much pleasure as collecting beetles. It was the mere passion for collecting, for I did not dissect them and rarely compared their external characters with published descriptions, but got them named anyhow. I will give a proof of my zeal: one day, on tearing off some old bark, I saw two rare beetles and seized one in each hand; then I saw a third and new kind, which I could not bear to lose, so that I popped the one which I held in my right hand into my mouth. Alas it ejected some intensely acrid fluid, which burnt my tongue so that I was forced to spit the beetle out, which was lost, as well as the third one.
>
> I was very successful in collecting and invented two new methods; I employed a labourer to scrape during the winter, moss off old trees and place [it] in a large bag, and likewise to collect the rubbish at the bottom of the barges in which reeds are brought from the fens, and thus I got some very rare species.[39]

Darwin's Cambridge friend Frederick Watkins (1808–1888), of Emmanuel College, later recalled:

> Many is the walk I have had with him in the meadows between Cambridge & Grantchester & many is the wretched animal that he unearthed from a rotten willow tree or some other obscure hiding place.
>
> I recollect that he introduced me to a beast wh. I have ever since held in holy hatred "Stapholinus olen" [Devils Coach Horse Beetle] wh. was for some time my idea of a scorpion & there was another the remembrance of whose name is "Crux major" or something like it.

After fifty years Watkins could still remember Darwin's enthusiasm for *Panagaeus cruxmajor*, or the crucifix ground beetle, one of Darwin's favourites. It occurs in Wicken and other fens under sedge litter. He later recalled "I am surprised what an indelible impression many of the beetles which I caught at Cambridge have left on my mind. I can remember the exact appearance of certain

[39] *Autobiography*, pp. 62–63.

posts, old trees and banks where I made a good capture. The pretty *Panagaus crux-major* was a treasure in those days".[40] Darwin's notes in his copy of *Systematic catalogue of British insects* (1829), p. 13, record his capture of one: "Cam: Spring 1828". In June 1828, Darwin reported excitedly to Fox "I Have taken Clivina Collaris, fig 3 Plate III of Stephens". Herbert recounted how one of the beetles was examined: "One day in his rooms he doped a large Beetle with a penful of prussic acid. The creature almost instantaneously kicked & fell over on his back apparently dead — but after a few minutes exposure to the lense it rallied and wattled off as if nothing had happened to it, to our general amazement."[41]

Darwin returned home to Shrewsbury for the summer, probably at the start of June. The College records books are strangely silent on his movements during this term. He visited friends and relatives for shooting and beetling. On 1 July 1828 he set out for Barmouth, Wales, for what he called his "Entomo-Mathematical expedition", a study holiday with a few other Cambridge undergraduates and their private tutors. And he named it in the appropriate order as he seems to have

E. Donovan, *The natural history of British insects* (1810) vol. 14., plate 477, fig. 2.

[40] *Autobiography*, p. 63.

[41] Herbert, J. M. 1882. [Recollections of Darwin, 12 June.] (DAR 112.A60-A61), transcribed by Kees Rookmaaker. *The Complete Work of Charles Darwin Online* (http://darwin-online.org.uk/).

Clivina collaris is listed in Stephens, *Mandibulata* 1: 40, plate iii, fig. 3.

done mostly entomology and little mathematics. Darwin's mathematical tutor "a very dull man" was George Ash Butterton (1805–1891) of St. John's.[42] Darwin confessed in a letter to Whitley: "I am idle as idle as can be: one of the causes you have hit on, viz irresolution the other is being made fully aware that my noddle is not capacious enough to retain or comprehend Mathematics. — Beetle hunting & such things I grieve to say, is my proper sphere".[43] Darwin wrote of this time in his *Autobiography*:

> I got on very slowly. The work was repugnant to me, chiefly from my not being able to see any meaning in the early steps in algebra. This impatience was very foolish, and in after years I have deeply regretted that I did not proceed far

[42] *Autobiography*, p. 58.

[43] To Charles Thomas Whitley [10 August 1828] [Barmouth], *Correspondence*, vol. 7:465.

enough at least to understand something of the great leading principles of mathematics; for men thus endowed seem to have an extra sense. But I do not believe that I should ever have succeeded beyond a very low grade.[44]

Darwin spent most of his time in Barmouth with two particular friends, John Maurice Herbert and Thomas Butler (1806–1886), the son of the grave headmaster at Shrewsbury School. What Butterton his tutor did while Darwin scrambled over mountains and popped beetles into bottles is not recorded. Herbert later described Darwin's activities:

taking our daily walk together among the hills behind Barmouth... Darwin entomologised most industriously, picking up creatures as he walked along, & bagging everything which seemed worthy of being pursued, or of further examination. And very soon he armed me with a bottle of alcohol, in which I had to drop any beetle which struck me as not of a common kind. I performed this duty with some diligence in my constitutional walks; but alas! my powers of discrimination seldom enabled me to secure a prize — the usual result on his examining the contents of my bottle, being an exclamation — "Well, old Cherbury" (the nickname he gave me, and by which he usually addressed me), "none of these will do." ... There is a remarkable rocky hill rising abruptly from the valley of the Dysynnie, called Craig Ackryer, or Bird's rock, which has swarms of Scabrids constantly frequenting its drags — Darwin went frequently there with his jars to secure any rare scabrid he cd. meet with; & he told me that he used to sit in a natural "chair" on the edge of the cliff, where he shot any bird on wing below him, which he wished to secure, and the guide who was at the foot of the cliff had to pick it up & carry it home for preserving.[45]

Butler later described the same holiday in Barmouth:

[Darwin's] own speciality however at that time was in the capture of beetles & moths & I learnt something of the voracity of the former seeing that when several were placed in the same box sometimes only one wd be found on our return the rest having been devoured: ... I remember also our killing the two largest vipers ... we buried them in order to catch some beetles which we found a week or more afterwards upon the spot[46]

[44] *Autobiography*, p. 58.

[45] Herbert, J. M. 1882. [Recollections of Darwin, 12 June.] (DAR 112.A60-A61), transcribed by Kees Rookmaaker. *The Complete Work of Charles Darwin Online* (http://darwin-online.org.uk/).

[46] Butler, Thomas. 1882. [Recollections of Darwin, 13 September] (DAR 112.A10-A12), transcribed by Kees

On 27 August Darwin left Barmouth early for the start of the shooting season on 1 September at Maer Hall, Staffordshire, the estate of his Wedgwood cousins seven miles from Etruria and Stoke-on-Trent where the pottery kilns churned out elegant Wedgwood ware. In September Darwin visited Fox at Osmaston Hall, near Derby.

Paley's Rooms?

Darwin returned to Cambridge for the Michaelmas Term on 31 October 1828.[47] Finally, there was a free room in College and the Tutor assigned Darwin to a comfortable set on the south side of First Court. There is a tradition that these rooms were once occupied by the famous natural theologian William Paley (1743–1805). This is not only given by Darwin biographers such as Janet Browne and Desmond and Moore, but has become one of those widely-believed tid-bits about Darwin's life and is mentioned by authors as various as the American biologist Jerry Coyne to the journalist and intellectual Christopher Hitchens.[48] Some writers even go so far as to state that Darwin was thrilled to be in Paley's old rooms. I have never seen any evidence that Darwin or anyone who knew him ever associated Darwin's rooms with Paley's. As far as I can discover, it is a very recent idea, perhaps only dating to the 1980s.

With the help of Geoffrey Thorndike Martin, I examined the College record books that record the room rents for Paley's undergraduate days at Christ's. Paley was admitted as a sizar on 16 November 1758 but began his residence in October 1759 at the age of sixteen. Between 1759 and 1764, the records reveal four different "Study rents". From the memorandum at the back of the record book, it is unclear to which rooms in College these rates refer.[49] I earlier attributed the idea

Rookmaaker. *The Complete Work of Charles Darwin Online* (http://darwin-online.org.uk/).

[47] There is a curious fragment of paper pinned to T.11.25 on the page for Xmas Quarter 1828 which on one side records "Table Cloth Bills 5.2.0. 127 to pay" and on the reverse reads:

Aegrotat Darwin

2. Nov. [1828?] John G[raham]

An Aegrotat was a medical certificate of illness. It is unclear why Darwin would require one on 2 November 1828 or for how long it may have extended. See the manuscript online here: http://darwin-online.org.uk/content/frameset?viewtype=side&itemID=CC-T.11.25&pageseq=4.

[48] Browne, Janet. 1995. *Charles Darwin*, vol. 1: Voyaging. London: Pimlico, p. 93; Desmond, Adrian and Moore, James. 1991. *Darwin*, p. 63; Coyne, Jerry A. 2009. *Why evolution is true*. Oxford: Oxford University Press, 2009; Christopher Hitchens, Arguably, 2011.

[49] "Study rents 1741–1782"

0 xmas quarter 1758?.

that Darwin and Paley lived in the rooms to College tradition. I now suspect that any such tradition has in fact derived from recent biographies, and not the other way around. Although the evidence is inconclusive, I think the story is most likely apocryphal.

When writing to his son William in 1858, Darwin described the exact location of his College rooms: "You are over the rooms which my cousin W. D. Fox had & in which I have spent many a pleasant hour. — I was in old court, middle stair-case, on right-hand on going into court, up one flight, right-hand door & capital rooms they were." William was at the time staying in what is now E6, though he later lived in his father's old rooms.[50] Darwin gave this lengthy description of the location because the College staircases were not then named with letters as they are today. Darwin's rooms are now known as G4, that is room 4 on staircase G. The rent was normally £4 per quarter.

3s Lady quarter, one of the lowest amounts.
3s Mids quarter
3s Mich quarter 1759 [12s per anum]
15s xmas quarter seems to be mid range price?
15s Lady quarter
15s Mids quarter
15s Mich quarter [£3 per anum] = rooms on chapel side and next to porters lodge
This rate remained the same until:
1.10.0 Xmas quarter 1760
1.10.0 Lady quarter [£6 per anum]
1.10.0 Mids quarter
1.10.0 Mich quarter
1.10.0 Xmas quarter 1761
1.10.0 Lady quarter 1762
1.10.0 Mids quarter – 6 people are charged this amount here.
1.10.0 Mich quarter 6 again.
1.10.0 Xmas quarter
1.10.0 Lady quarter 1763 [Paley earned his BA in 1763.]
3s Mids quarter
3s Mich quarter
3s Xmas quarter
3s Lady quarter
3s Mids quarter 1764
3s Mich quarter
3s Xmas quarter
Notes on study rents at the back of the volume.
1.10.0 corner room by Master & gate
Ground rooms were the same?
£3 per anum on chapel side and next to porters lodge.
£4 per anum 5 rooms on hall side.
£4 per anum Middle chambers in old court.

[50] F. Darwin 1914.

William Paley. Mezzotint by John Jones, after George Romney, 1792.

Darwin's rooms consisted of a panelled main sitting room (c. 8 x 8m) with an adjoining dressing room and bedchamber. (The rooms and their restoration are described in greater detail below.) His three windows on the north overlooked First Court, with the Chapel directly across from him. His postulated clerical future was never out of view. The Master's Lodge was to its right and closer still the Hall. Darwin's south facing windows overlooked Bath Court, what is today the site of the new undergraduate library.

"The first court, Christ College." Engraving showing the view from just under Darwin's windows in the 1830s.

Worldly study

Imposition

First term/ second term

[study/ riding]

A set of humorous etchings depicting the decline of university students in 1829 from initial study into the vices of drinking, smoking and riding. William Makepeace Thackeray, [1878.] *Etchings by the late William Makepeace Thackeray while at Cambridge: illustrative of university life, etc., etc. Now printed from the original plates.* [London: H. Sotheran and Co.].

It was possible to buy the furniture left by the previous tenant from the College upholsterer. Darwin may have done so. It was common to buy crockery, tea sets, decanters and wine glasses from the bedmaker who traditionally inherited any left by the previous occupant. A large bill of £40 5*s* 6*d* for a woollen draper in the Easter term may have been for a carpet ordered for his new rooms.[51] In December Darwin wrote to his brother Erasmus: "After you left Cambridge. I got into very nice rooms in College, far more comfortable than lodgings, as you will find when you come next to Cambridge. I imbibed your tastes about prints, and put it into practice, and have bought some very good prints, which I long for you to see."[52]

A college servant, known in Cambridge as a gyp, was assigned to each staircase. The word gyp is from the Greek for vulture, though by Darwin's time the reason why gyps were thus named was lost in obscurity. Darwin's gyp is recorded simply as 'Impey'. Gyps delivered letters, brushed clothes, ran errands and made coffee. The College porters took letters to and from the post office. Another servant known as a shoeblack cleaned and polished shoes and boots. These services were listed separately in the College accounts for each student. For example, in the quarter ending Lady Day (25 March) 1830 Darwin was charged £4.12.6 for coal, 7 shillings for the shoeblack and £2 one shilling for the barber.

A typical day at Christ's College for Darwin probably began around 7am when he was awoken by Impey in time to dress for Chapel at 7.30am. While attending Chapel the bedmaker, apparently a Mrs. Field, would come in and make the bed. This too was charged on the account, in this case £1 one shilling for the same quarter. Darwin would return from Chapel and have breakfast in his rooms before a blazing coal fire with the kettle boiling on the grate. The table was laid by Impey and the College breakfast brought from the kitchens, according to the recollection of a near contemporary at Trinity, consisted of: "the fourth part of a half-quartern loaf, and twopenny-worth of butter".[53] Tea and coffee and any other extras were provided or paid for by the student. After breakfast Impey cleared the breakfast things. It was certainly a more luxurious lifestyle than any university student nowadays could imagine. But in Darwin's day higher education was usually for the social elite and with cheap labour and a frankly and unashamedly hierarchical society, such pampering was to be expected.

---

[51] T.11.26.

[52] *Correspondence*, vol. 1: 71.

[53] [Atkinson, S.] 1825. Struggles of a poor student through Cambridge. *London Magazine and Review* (new series) 1, p. 507.

Clearly a wide variety of activities took place in Darwin's rooms. He read for his College curriculum, wrote letters, compared his captured beetles with published descriptions and carefully pinned his captured beetles to cork boards. Stephens was one of the new generation of entomologists who ensured that their science moved away from mere collecting for show and display and became more of a well-organized science with species carefully described and delineated under the system of Linnaeus. He had friends to coffee, and in the evenings they sometimes dined there and would then drink wine and play cards. He must have spent many hours curled up in one of his spacious bow window seats reading a book.

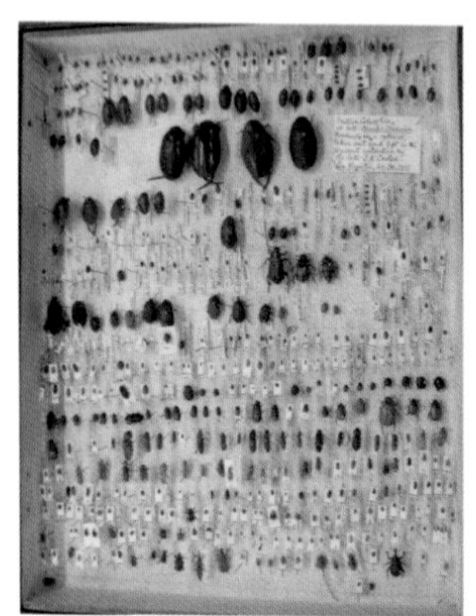

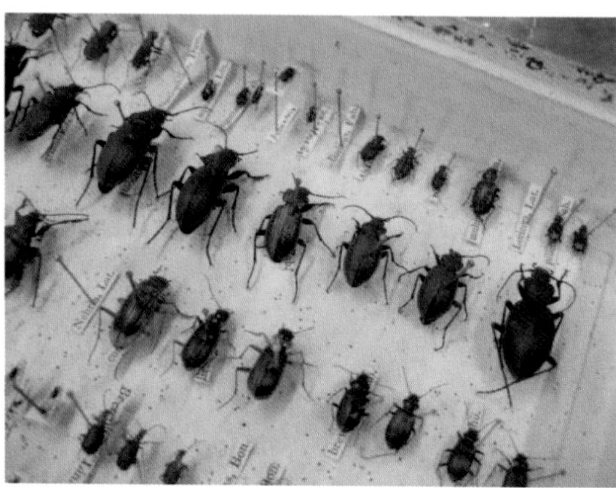

Some of Darwin's Cambridge beetles, (transferred from an original cabinet) now at the Zoology Museum, Cambridge.[54]

Another Cambridge friend, Jonathan Henry Lovett Cameron (1807–1888) from Trinity College, recalled: "At Cambridge I used to read Shakspere with him in his own room & he took great pleasure in these readings. He was also very fond of music, though not a performer & I generally got an order for him for Kings Coll. Chapel on Sunday evening." By the end of his life, however, Darwin complained that "I have tried lately to read Shakespeare, and found it so intolerably dull that it nauseated me."[55] But of music in Cambridge, he recalled: "I used generally to go by myself to King's College, and I sometimes hired the chorister boys to sing in my rooms."[56] Herbert remembered accompanying Darwin to the

[54]The former photograph was gratifyingly reproduced in Attenborough, David. 2009. *Life stories*. London: HarperCollins.

[55]*Autobiography*, p. 138.

[56]*Autobiography*, p. 61.

afternoon service at Trinity "when we heard a very beautiful anthem — At the end of one of the parts, which was exceedingly impressive, he turned round to me & said with a deep sigh How's your backbone?" [57]

Long before coming up to Cambridge Darwin was a passionate sportsman. In October 1828, Darwin's father and sisters contributed £20 towards the purchase of a new double barrelled shotgun. With it, Darwin often went shooting in the surrounding fens. He also acquired a dog, named Dash to accompany him. Even the generation before the Christ's student Henry Gunning recalled: "The great source of idleness, which consumed more time than all my other employments put together, was my passion for shooting, for which diversion Cambridge afforded the most extraordinary facilities."[58] When Darwin could not go shooting, he would practice in his rooms as he later recalled:

> When at Cambridge I used to practise throwing up my gun to my shoulder before a looking-glass to see that I threw it up straight. Another and better plan was to get a friend to wave about a lighted candle, and then to fire at it with a cap on the nipple, and if the aim was accurate the little puff of air would blow out the candle. The explosion of the cap caused a sharp crack, and I was told that the Tutor of the College remarked, "What an extraordinary thing it is, Mr Darwin seems to spend hours in cracking a horse-whip in his room, for I often hear the crack when I pass under his windows."[59]

About this time, Herbert recalled that he and Darwin had an "earnest conversation about going into Holy Orders; & I remember his asking me with reference to the question put by the Bishop in the Aduration service: 'Do you trust that you are inwardly moved by the Holy Spirit &c' whether I could answer in the affirmative — : & on my saying 'I could not,' he said 'neither can I, & therefore...I can not take orders.'"[60] This recollection may either be not entirely accurate or Darwin may later have resolved to go through with ordination anyway. In a letter from May 1830, Darwin wrote to Fox: "I have some thoughts of reading divinity with [J.S. Henslow] the summer after next."[61]

---

[57] Herbert, J. M. 1882. [Recollections of Darwin, 12 June.] (DAR 112.A60-A61), transcribed by Kees Rookmaaker. *The Complete Work of Charles Darwin Online* (http://darwin-online.org.uk/).

[58] Winstanley, D. A. ed. 1932. *Henry Gunning: Reminiscences of Cambridge*. Cambridge: University Press, p. 35.

[59] *Autobiography*, p. 44.

[60] Herbert, J. M. 1882. [Recollections of Darwin, 12 June.] (DAR 112.A60-A61), transcribed by Kees Rookmaaker. *The Complete Work of Charles Darwin Online* (http://darwin-online.org.uk/).

[61] *Correspondence*, vol. 1: 104.

Great St Mary's Church Exterior. 1841. T. D. Atkinson. *Cambridge described and illustrated.* Macmillan, 1897.

Chapel of Christ's College. R. Ackermann, *History of Cambridge, 1815.* From the 1909 Shipley album in the Old Library. This print was part (item 131) of the 1909 Darwin exhibition at Christ's.

Members of College were called to Chapel by the bell, which rang for five minutes as the start of services approached at 7.30am. Students were probably required to attend Chapel eight times per week, at least one service per day and twice on Sunday. The former Chaplain of Christ's, the Rev. Christopher Woods, helpfully provided the following remarks.

> It appears that when the bell stopped ringing, late-comers were not permitted to enter. Fines were payable by those who missed chapel. The chapel calendar was, of course, based upon the Christian religious calendar, of those holy days and saints' days which were maintained by the Church of England after the Reformation. According to the College statutes there would have been divine service of Morning and Evening Prayer (Matins and Evensong) on normal ('ferial') days at 7.30am and 6.00pm. On Sundays, chapel would have been at 9.30am, immediately preceding the University Sermon at the University Church of St Mary the Great. Attendance at the University Sermon was a weekly exercise, when scholars and students would have been exposed to the most contemporary and often fervent Christian polemic. As such, Cambridge students would not have been ill-educated in public theology or Philosophy of Religion. On College commemorations or holy days and feast days (which appeared in the *Cambridge Calendar*, as they do to this day), divine worship would have been at 3.30pm. Surplices (long white garments still used today by clergy and choristers), were worn by all Scholars and Fellows on Sundays, Saturday evenings, morning and evenings of feast days and on the evening before the same feast days. Surplices would have been kept in Chapel for those in College.
>
> Because of the puritan nature of Christ's College, by dint of history, ceremonial would have been minimal and ideas of reformed theology would have been current and commonplace. As a College of the University, bound by statute of the Established Church, the Celebration of the Holy Communion (Sacrament of the Lord's Supper) was obligatory for all at certain times of the year. To this day, we would know these days as 'scarlet days' (days when doctors of the University are obliged to wear festal gowns in public): Christmas Day, Easter Day, Whitsunday, possibly Trinity Sunday and also on the Sundays after the division of term in the Michaelmas and Lent terms, the Second Sunday in Easter Term and the second Sunday in August.

Undergraduates probably rotated through readings in Chapel. It is interesting to note, and no biography has ever done so, that Charles Darwin, although he may not have been the most "distinct and graveful reader" almost certainly stood at the medieval brass lectern in Christ's College Chapel and read from the Bible. It is now a striking and paradoxical image. Charles Darwin is probably the man of science who has been attacked above all others in history as irreligious or dangerous to religion.

Far from being an establishment-defending and entrenched Anglican stronghold Christ's had a large percentage of reform-minded Fellows. "In 1837 a draft of a new body of statutes was made, and on 24 February 1838 an order was signed by the Master and ten Fellows (including Shaw, Baines, and J. Cartmell, the future Master) that a petition for the substitution of [the Elizabethan] Statutes should be presented to the Queen. … [the draft statutes] seem to have permitted the marriage of Fellows, and some participation in College emoluments by persons not members of the Church of England. …Divine service should be held on Sundays only in the College chapel…" and so forth.[62] However, the draft was vetoed by the Visitor. Francis Darwin recorded a recollection of his father from his student days in Cambridge showing how lax the religious atmosphere of the College was in his time:

> The way in which the service was conducted in chapel shows that the Dean, at least, was not over zealous. I have heard my father tell how at evening chapel the Dean used to read alternate verses of the Psalms, without making even a pretence of waiting for the congregation to take their share. And when the Lesson was a lengthy one, he would rise and go on with the Canticles after the scholar had read fifteen or twenty verses.[63]

Darwin's tutors, Joseph Shaw and John Graham, are buried in the antechapel. Their Latin-inscribed tombstones in the floor can still be read today. There is also a memorial to Shaw in the Chapel.

College Lectures

After breakfast in their rooms students were expected to attend the two College lectures in the lecture room, probably from 9–11am. Darwin recalled "With respect to Classics I did nothing except attend a few compulsory college lectures, and the attendance was almost nominal."[64] According to a work pub-

---

[62] Peile, J. 1900. *Christ's College*. London: Robinson, p. 279.

[63] Darwin, Francis. (ed.) 1887. *The life and letters of Charles Darwin*. London: Murray, vol. 1: 165.

[64] *Autobiography*, p. 58

lished in 1830, *Classical examinations; or, a selection of University scholarship and other public examination papers, and of the question papers on the lecture subjects of the different Colleges in the University of Cambridge*:

> The principal Classical Lectures are in most of the Colleges given to men in their first year of residence. The subjects are greatly varied; but they are more usually a Greek play, a selection from one of the Greek historians, orators, or philosophers, and from some Latin writer, either of prose or verse. … [in some colleges] a portion of the Greek Testament, one of the Gospels, or the Acts of the Apostles, is a subject of examination for second-year men. And besides this, some classical book is read in the lecture-room, generally one of those which form a part of the previous Examination, or *Little Go*. The first and second terms of the second year, and not unfrequently also the second and third of the first year are employed by the tutors of the smaller Colleges in preparing their men for this public examination. [For the third year, undergraduates are] kept employed upon the six first books of the Iliad and Æneid, which form the classical part of the examination for ordinary degrees.[65]

CHRIST COLL. 1828.

1. In what places of an Iambic Trimeter do the tragic poets admit Anapæsts, Dactyls, and Spondees? What exception is there to the rule concerning Anapæsts? And what limitations to the admissibility of a Spondee in the fifth place of the verse?

2. State the rules applicable to a regular system of Anapæstic Dimeters. What is meant by the term *συνάφεια*?

3. Translate v. 258—73.

ἀτὰρ τί δὴ σόφισμα τοῦθ' ἡγούμενοι

. . . . .

ἄκουσον.

Explain the syntax of v. 260.

A classics examination paper (College Examination on Euripides, Hecuba) at Christ's College from 1828. (Anon 1830)

[65] See also [Atkinson, S.] 1825. Struggles of a poor student through Cambridge. *London Magazine and Review* (new series) 1: 491–510.

CONIC SECTIONS.

CHRIST'S COLLEGE, *May* 1829.

1. DEFINE the parabola and the ellipse.

2. A tangent at any point P in a parabola bisects the angle SPM.

3. The sub-tangent is double of the abscissa AN.

4. Find the value of the perpendicular from the focus on the tangent to any point in the parabola, and in the same curve shew that $Sy \propto \sqrt{SP}$.

5. In the parabola $4\,SP \,.\, PV = QV^2$.

6. In two parabolas about the same axis, having a common ordinate QPN, the area ASP : area ASQ :: $\sqrt{L} : \sqrt{L'}$, where L, L' are the latera recta of the two parabolas.

7. In an ellipse the centre bisects all diameters.

8. Perpendiculars from the foci upon the tangent at any point in the ellipse, meet the tangent in the circumference of a circle, whose diameter is the axis major.

9. The sub-normal $NG = \frac{L}{2\,AC} \cdot CN$, where L = latus rectum of the ellipse.

10. $PV \,.\, VG : QV^2 :: CP^2 : CD^2$.

11. Find the diameter and chord of curvature to an ellipse through the centre.

12. Shew that the ellipse is a conic section.

A mathematics examination paper used at Christ's College in May 1829, when Darwin was a student.

A number of College examination papers in mathesmatics from Darwin's time have recently been discovered at Christ's.

Dr Robert Hunt, Lecturer in the Department of Applied Mathematics and Theoretical Physics and Fellow of Christ's, provided the following comments on the sample mathematics paper reproduced above:

> Conic sections are the shapes that you get when a flat plane intersects with a cone: they were first studied extensively by the Greeks, who showed that you always get a circle, ellipse, parabola or hyperbola (except in a few very special and uninteresting cases). The Greeks took a particular interest in geometry, including geometry in three dimensions, for purely intellectual satisfaction.
>
> In the seventeenth century conic sections acquired a new significance when, amongst his many other discoveries such as the calculus, Isaac Newton showed that heavenly bodies (planets, comets, etc.) always move through the heavens under the influence of the Sun's gravity in the shape of a conic section. He was therefore able to demonstrate using his Universal Law of Gravitation that the Earth moves around the Sun in an elliptical shape, while comets within the Solar System typically move either in the shape of a parabola or that of an ellipse. Edmund Halley's use of this theory to predict the return of the comet now named after him was a decisive vindication of Newton's work.
>
> Newton's influence within Cambridge in particular and England in general meant that the serious study of conic sections (both in the abstract and with applications to planetary motion) was still a major part of the Cambridge curriculum in Darwin's time. In Continental Europe, however, mathematicians had moved forward in innovative directions, creating whole new areas of mathematics (such as analysis, statistics and the calculus of variations) that were to prove key to future scientific discovery, while in England the syllabus was held back with a large amount of recycled, old fashioned material. It was not until the late nineteenth century that English mathematics began to absorb many of the new, modern concepts from Europe.

After College lectures, students might have tea or coffee in their rooms and were free to visit friends or go for walks. Around 1 o'clock, it was customary to visit private tutors. Dinner was then at 4 o'clock in Hall. For Pensioners like Darwin, the standard meal or commons provided was essentially still Elizabethan fair which consisted only of joints of meat and beer. However to keep up with the expectations of the times, extras could be purchased such as vegetables, pies and cheese. This is why the College record book T.11.25 had a separate column for vegetables next to the record for commons. In 1828, vegetables were charged at "5 ½ per day". In the first quarter of Darwin's residence in Cambridge his vegetable bill came to £1 *2s* 5½*d.* (T.11.25)

The Hall

From a photograph by]
THE HALL
[J. Clarke Palmer, Cambridge
Plate VI

The Hall of Christ's College, c. 1900 from Peile 1900. Note the plaster ceiling which has since been removed to expose the wooden beams.

The Hall was the largest building in Christ's, and the place where members of the College gathered for dinner. The Hall today is quite different from its appearance in Darwin's day. "In 1723 money was given to 'beautify' it: which was done by putting deal wainscot over the oak panelling, by covering up the old fireplace, and by hiding the roof with a cylindrical plaster ceiling." It remained thus until 1875 when "the old roof was removed, reconstructed, and replaced; the walls were rebuilt and raised about 6 feet...and new oak panelling was put up, some of the original work being sufficiently sound to remain at the south end." The panelling at the north end was added in 1900.[66] The two little windows which allowed the Foundress to observe the hall from her apartments, long plastered or painted over, were reopened in 1905.

Before the meal a Latin grace was read.[67] The Fellows and Fellow Commoners ate a three course meal with port or sherry at their own table on a raised dais at the north end of the Hall. The Pensioners sat together at long tables at the south end of the Hall. The Sizars acted as waiters. After dinner, the Fellows and Fellow Commoners would retire to the Combination Room located upstairs behind the gallery at the south end of the Hall. Here, they would drink wine or

[66] Peile, J. 1900. Christ's College. London: Robinson, p. 31.

[67] Exhiliarator omnium Christe
Sine quo nihil suave,
nihil jucundum est:
Benedic, quaesumus, cibo et potui
servorum tuorum,
Quae jam ad alimoniam corporis
apparavisti;
et concede ut istis muneribus tuis ad
laudem tuam utamur
gratisque animis fruamur;
utque quemadmodum corpus nostrum
cibis corporalibus fovetur,
ita mens nostra spirituali verbi tui
nutrimento pascatur
Per te Dominum nostrum.

Christ, the gladdener of all,
Without whom nothing is sweet, nothing
pleasant:
Bless, we beseech you, the food and drink of
your servants,
Which you have now provided for the
nourishment of the body;
And grant that we may use these gifts of
yours for your praise,
And enjoy them with grateful minds;
And that, just as our body is nourished by
bodily foods,
So our mind may feed on the spiritual
nourishment of your Word.
Through you, our Lord.

port and converse until evening Chapel at 6pm. This gathering was and is known as "the room" to whose continuing traditions and secrets only the Master and Fellows of the College are initiated.

The Library

From a photograph by] THE LIBRARY [*J. Palmer Clarke, Cambridge*
Plate VII

The Old Library of Christ's College from Peile, J. 1900. *Christ's College*. London.

The library, essentially what is today called the Old Library, held all the books of the College until 1852. Here Darwin was able to consult the books needed for his College studies. The University Library (formerly called the Public Library) was housed in part of what are now the Old Schools next to Senate

House. Ancient Greek and Roman statues and other priceless antiquities were on display — there was even an Egyptian mummy and a copy of the famous Rosetta stone. The library boasted 100,000 books and 2,000 manuscripts of many languages and epochs. It was even than [then] a copyright library, meaning that it was one of the few libraries in the country to be entitled to one copy of every book published. In addition to its endowments, every member of the university was required to contribute 6 shillings to the library per annum. But it was at this time not open to the use of undergraduates.

On 20 December 1828, Darwin arrived back in Shrewsbury with his new dog Dash. He visited friends and relatives in Shropshire and Staffordshire for shooting and social calls. In a January 1829 letter to Fox, Darwin made a rare reference to his College studies: "my Studies consist of Adam Smith & Locke, in the latter of which I suppose you are an adept, & I hope you properly admire it — About the little Go I am in doubt & tribulation. I have had very little shooting."[68] "Little Go" was student slang for the University's Previous Examination, which undergraduates took in their second year. "The subjects of examination are one of the four Gospels or the Acts of the Apostles in the original Greek, Paley's Evidences of Christianity, one of the Greek and one of the Latin Classics".[69] Locke's *An essay concerning human understanding* (1690) appeared in the B.A. examination, known as the Poll, for those who were not candidates for honours, known as the Tripos. The work by Adam Smith, probably his *Theory of Moral Sentiments* (1759), did not appear in the Previous or B.A. Examinations. It may have been a reading assigned by his College Tutor or Lecturer. Before returning to Cambridge for Lent Term Darwin stayed in London a few days with his brother Erasmus and visited one of the main players in entomology, Frederick William Hope, just a few years Darwin's senior.

[68] *Correspondence*, vol. 1: 74.

[69] *Cambridge University Calendar*, 1829, p. 169.

SECOND YEAR AT CAMBRIDGE

Darwin returned to Christ's on 24 February 1829. Fox was gone, he had taken his B.A. degree on 23 January, ranking 88th out of 160. Earlier in the month, Darwin's cousin Hensleigh Wedgwood was elected a Finch and Baines Fellow, but he would hold this only until October 1830. Two days after returning to his comfortable rooms, Darwin wrote Fox to report on his stay in London where he had visited entomologists F. W. Hope and James Francis Stephens. The former had generously given Darwin specimens of 160 beetle species for his collection. Darwin ordered a beetle cabinet to help house his growing collection. Darwin also reported that "By Grahams decided advice, I do not go in for my little Go." John Graham (1794–1865) replaced Shaw as Tutor in 1829. He clearly did not think Darwin was ready to pass the examination.[1] Shipley described Graham as "one of the most brilliant of the *alumni* of the College (fourth Wrangler and Chancellor's Classical Medallist in 1816), who was elected Master of the College in 1830, and was appointed to the Bishopric of Chester in 1848. Graham was one of the small group of Cambridge Liberals in the days of the first Reform Bill,

[1] S. M. Walters and E. A. Stow, *Darwin's Mentor.* 2001, p. 79 suggest Graham saw Darwin as "a typically idle 'poll man' destined for the Church, dawdling away his Cambridge days with his horse and his gun until thrown into a panic by the approaching examinations." This quotation, however, is from E. S. Leedham-Green, *A concise history of the University of Cambridge* 1996, p. 141 and is not based on any evidence from Graham.

and a strong supporter of the abolition of University tests. As a disciplinarian in College, he is said to have been somewhat too "easy-going".[2] After Graham was elected Master in 1830, Darwin's Tutor was Edward John Ash (1799–1851).

Darwin may well have benefited from this lack of discipline. He wrote to Fox on 1 April 1829: "Last night there was a terrible fire at Linton eleven miles from Cambridge; seeing the reflection so plainly in sky, Hall, Woodyeare Turner & myself thought we would ride & see it we set out at 1/2 after 9, & rode like incarnate devils there, & did not return till 2 in the morning. altogether it was a most awful sight."[3] Returning this late should technically have incurred punishment, but there is no record of any disciplinary consequences. Either the College took no notice, as when Fox returned "rather late" or perhaps Darwin climbed over the wall as Fox did when returning once at 4.30 am.[4]

With his close friend and fellow entomologist Fox no longer present, Darwin spent more time with other friends. One of these was an old Shrewsbury school friend, John Price (1803–1887), who remembered: "we were walking up the chalk path to Cherry Hinton Quarries. … [Darwin] halted with a characteristic & expressive stamp, exclaiming 'Price Price, what w.d I give to be such a naturalist as you' !! This rhapsody will be more amusing, if I add that I believe it was evoked by the simple fact of my knowing Yarrow & other common plants equally well in winter."[5]

On 15 March, Darwin wrote to Fox of the routine in his College rooms: "I am leading a quiet every-day sort of a life; a little of Gibbons history in the morning & a good deal of *Van. John* [Blackjack], in the evening this with an occasional ride with Simcox & constitutional with Whitley, makes up the regular routine of my days." On 1 April, Darwin reported to Fox a confrontation resulting from competition with another Cambridge undergraduate who was raiding Darwin's beetle supply:

> Entomology goes on but poorly: a few Dromius & Agonum's, together with the Pæcilus (with red thighs) make the g[reat] part of what I have collected this ter[m]. I have caught Mr. Harbour letting [Charles Cardale] Babington have the first pick of the beetles; accordingly we have made our final adieus, my part in

---

[2] Shipley, A.E. [1924]. *Cambridge cameos*, p. 127.

[3] *Correspondence*, vol. 1: 81.

[4] Diary 29 Nov. 1824; Diary 23 Nov. 1825.

[5] [Price, John.] nd. [Recollections of Darwin.] (DAR 112.B101-B117), transcribed by Kees Rookmaaker. *The Complete Work of Charles Darwin Online* (http://darwin-online.org.uk/).

> the affecting scene consisted in telling him he was a d—d rascal, & signifying I should kick him down the stairs if ever he appeared in my rooms again[6]

Darwin never shed his dislike for Babington.[7]

A very interesting and dramatic episode occurred while Darwin was in Cambridge as first revealed in the biography of Darwin by Desmond and Moore. The radical freethinkers Richard Carlile (1790–1843) and the Revd Robert Taylor (1784–1844) began their "infidel missionary tour" in Cambridge in May 1829. They challenged the Vice Chancellor, the senior divines and the Masters of the Colleges to debate the validity of Christianity.

> The Rev. Robert Taylor, A.B., of Carey-street, Lincoln's Inn, and Mr. Richard Carlile, of Fleet-street, London, present their compliments as Infidel missionaries, to (*as it may be*) and most respectfully and earnestly invite discussion on the merits of the Christian religion, which they argumentatively challenge, in the confidence of their competence to prove, that such a person as Jesus Christ, alleged to have been of Nazareth, never existed; and that the Christian religion had no such origin as has been pretended; neither is it in an way beneficial to mankind; but that it is nothing more than an emanation from the ancient Pagan religion. … They also impugn the honesty of a continued preaching, while discussion is challenged on the whole of the merits of the Christian religion.[8]

All was reported in (one-sided) triumphant tones in Carlile's journal *The Lion*. The University authorities ignored the radicals except for spitefully withdrawing the license of the innocent man who owned their lodgings. The whole affair must have sparked a great deal of conversation in the colleges.

Desmond and Moore used this episode to claim: "In later years [Darwin] would remember Taylor as 'the Devil's Chaplain', fearing that he himself might be similarly reviled, an outcast from respectable society, a terror to the innocent an infidel in disguise." This is based on a famous passage in a letter by Darwin to his friend the botanist J.D. Hooker in 1856: "What a book a Devil's chaplain might write on the clumsy, wasteful, blundering low & horridly cruel works of nature!"[9]

---

[6] *Correspondence*, vol. 1: 81.

[7] See for example to J.D. Hooker, 18 [May 1861] *Correspondence*, vol. 9: 133.

[8] *The Lion*, no. 21. vol. 3, 1829. p. 641.

[9] Darwin to J. D. Hooker 13 July [1856], *Correspondence*, vol. 6: 178.

There are several serious problems with their claim. First of all the statement means that contrary to a pious clergyman praising the beauty and beneficence of nature, a non-believer could make an extraordinary case on the basis of the many examples of cruelty in nature. Secondly, we have no evidence of any kind that Darwin was even aware of the visit of Taylor and Carlile. Although he probably heard something about their visit, we know nothing about any thoughts he may have had, if he cared to have any opinions at all. Secondly, Taylor was only dubbed "the devil's chaplain" two years *after* his visit to Cambridge, during a series of extremely radical lectures at the London Blackfriars Rotunda in 1831. Thirdly, the phrase "devils' chaplain" goes back at least to Chaucer. By the time Darwin wrote this letter, the phrase had appeared hundreds of times in books and periodicals — thus it was a well-known figure of speech. Hence, there is no link in the phrase used by Darwin in 1856 to the events in Cambridge in 1829. And finally, the attribution of fear of being reviled by respectable society, although widely believed, is based on no evidence.[10]

"Pembrok Hall & c. from a Window at Peterhouse". R. Ackermann's, *History of Cambridge*, 1815.

[10] See Wyhe, J. van. 2007. Mind the gap: Did Darwin avoid publishing his theory for many years? *Notes and Records of the Royal Society* 61: 177–205 and Wyhe, J. van. 2013. *Dispelling the Darkness: Voyage in the Malay Archipelago and the discovery of evolution by Wallace and Darwin.* Singapore: World Scientific Press. The original context of the letter, in fact, might allow one to conclude that Darwin meant that T.H. Huxley would make a devil's chaplain. Despite this many writers continue to assert, erroneously, that Darwin referred to himself as a devil's chaplain. See also Corsi, P. 1998: A Devil's Chaplain's calling? *Journal of Victorian Culture* 3(1): 129–137.

In June Darwin may have received the latest number of Stephens' *Illustrations of British entomology*, Haustellata vol. 2 (appendix) dated 1 June 1829. On page 200 appeared the record of Darwin's capture of *Graphiphora plecta* (the flame shoulder moth, now *Ochropleura plecta*) and the first word Darwin ever published: "Cambridge", the location of capture. On the 15th another number of Stephens appeared, this time with thirteen species of beetle collected by Darwin. He later described the feeling of seeing his captures in print: "No poet ever felt more delight at seeing his first poem published".[11] It has sometimes been pointed out that this exact wording was not printed in Stephens. But the entry in *Mandibulata* (volume 3, p. 266) states: "captured by the Rev. F. W. Hope and C. Darwin, Esq., in North Wales".[12]

H.N. Humphreys & J.O. Westwood, *British Moths and Their Transformations*. London: William Smith, vol. 1, 1845, Plate 26, fig. 7. Source: Kurt Stüber, www.BioLib.de

Page 136. Graphiphora plecta. "Cambridge."—*C. Darwin, Esq.*

Darwin's first published word was "Cambridge".

No complete list of Darwin's beetle collection seems ever to have existed. Four principal textual sources survive that reveal details of his collection. For a time

---

[11] *Autobiography* p. 63.

[12] Stephens, J.S. 1830. *Illustrations of British entomology*: Mandibulata vol. 3, p. 266. With thanks to J. David Archibald for informing me of this reference.
Correspondence, vol. 1:98: "The report of the moth, if it appeared early in June, would be the first time CD's name appeared in print, but in the Autobiography, p. 63, where he writes of his delight at first seeing the magic words, "captured by C. Darwin, Esq.", his memory is of beetle collecting. Unfortunately the 'magic words' do not help in making a clear decision about the date or the publication; they do not occur in either list, nor have they been found elsewhere in Illustrations of British entomology".
Darwin's name first appeared in print in Grant, R.E.. 1827. Notice regarding the ova of the Pontobdella muricata, Lam. *Edinburgh Journal of Science* 7 July: 160–161: "The merit of having first ascertained them [ova] to belong to that animal is due to my zealous young friend Mr Charles Darwin of Shrewsbury, who kindly presented me with specimens of the ova exhibiting the animal in different stages of maturity."

between 1828 and the summer of 1829, Darwin kept a list of 56 beetle species in his old *Edinburgh notebook*. He also made annotations next to the descriptions of 281 species in Stephens, *A systematic catalogue of British insects* (1829) — sometimes recording where he collected a species with the date or the person who gave him a specimen. Darwin's copy is in Cambridge University Library. Thirty-two species were detailed as captured by "C. Darwin Esq." in Stephens, *Illustrations of British entomology* (1827–1845). And finally Darwin's 1828–1830 letters to Fox mentioned 53 species. He obviously became more proficient in capturing and identifying beetles over the years. These records combined reveal 390 species (sometimes genus only).[13]

The list from the *Edinburgh notebook* (ff. 107v–105v) is below, transcribed for me by Kees Rookmaaker:

No.			
1	Harpalus	Tardus	
2	. . . . .	rubripes (2 specimens) male & female	[1–3] Southend
3	. . . .	rufimanus	
4	*[Bryphillus.]* [=? Biphyllus]		Gloucestershire
5	Hydroporus	lineatus	
6	. . . . .	picipes	[5–6] Southend
7	Berosus	aericeps	
8	Dermestes	vulpinus	
9	Omaseus	anthracinus ?	Hope
10	Hydroporus	humeralis	
11	Harpalus	anxius	
12	[Ptinus] Fur	[Ptinus] lichenum.	Mr Waterhouse
13	Donacia	proteus.	Waterhouse
14	Philochthus	biguttatum	
15	Helophorus	~~tri~~ viridi collis	
16	Hydrobius	substriatus.	Hope
17	Cassida	ferruginea.	

(*Continued*)

[13] The definitive study of Darwin's insects is Smith, Kenneth G. V. 1987. Darwin's insects: Charles Darwin's entomological notes, with an introduction and comments by Kenneth G. V. Smith. *Bulletin of the British Museum Natural History) Historical Series*. Vol. 141): 1–143. Smith was unaware of the list in DAR118 which is provided here. For further information see http://darwin-online.org.uk/EditorialIntroductions/vanWyhe_Stephens.html.

(*Continued*)

18	Tarus	basalis.	
19	Panagaeus	4 pustulatus.	(Hope)
20	Hydroporus	areolatus	
21	Dromius.	fasciatus.	Southend
22	Rhyzobius		
23	Cryptophagus		

32. Cryp: always taken on the sea shore near or upon a creeping sort of wild rose. — early in July . — 29.
33. Cryp: on Barmouth rocks. August. — 29
36. Nitid. on old bones. Summer. — 29
38. Rhy: taken together with some Nitidulae in a wild Bees nest. June. — 29
39 & 40. Anth: taken chiefly on umbelliforous plants.
41. Ips. May. 1829

No.			
24	Scolytus	destructor	Cam
25	Engis	rufifrons	
26	Gryphus	equiseti	
27	Cateretes		
28	Notaphus		
29	Haliplus	obliquus.	Cam
30	Lopha	poecila.	Cam
31	Silvanus.		Cam:
32	Cryptocephalus	minuta.	Barmouth
33	Cryp: ...	bipustulatus	Do
34	Cryp.	gracilis.?	Weaver
35	Nitidula	punctatissima.	Woodhouse June. 29
36	. . . . .	sordia.	Shrewsbury
37	Meligethes	rufipes	(rose blossoms)
38	Rhyzophagus	. . .	
39	Anthemus	varius.	Shrew.
40	An: . .	verbasii	Shrew.
41	Ips	ferruginea?	Shrew. (taken flying)
42	Byrrhus	dorsalis.	Gravel pits May. — 28
43	Malachius	ruficollis.	Cam. Summer. — 29
44	Elater	segetis.	Passim
45	Hylesinus	sulcatus.	June. Shrew: — 29

(*Continued*)

(*Continued*)

46	Callidium	bagulum.	Hope
47	Dromius	bifasciatus	Weaver
49.	Het. creeping in great profusion together with laevigatus. & Cillenum laterale on a muddy bank which was daily overflowed by the tide. We only found it on the opposite side of the river to Barmouth		
51.	Gal: in great profusion		

No.			
48	Ocys	melanocephalus	Cam. Summ: — 29
49	Heterocerus	obsoletus.	Barmouth
50	Callidium	alni.	Kent
51	Galeruca	lineola.	Barmouth. July. — 29
52	Gal.	crataegi.	Cam. — 28
53	Crysomela[=Chrysomela]	quinquejugis.	Barmouth
54	Cercyon	quisquilius.	Shrews:
55	Platysma	niger	
56	Chaetophora?	—	Maer. stone quarries

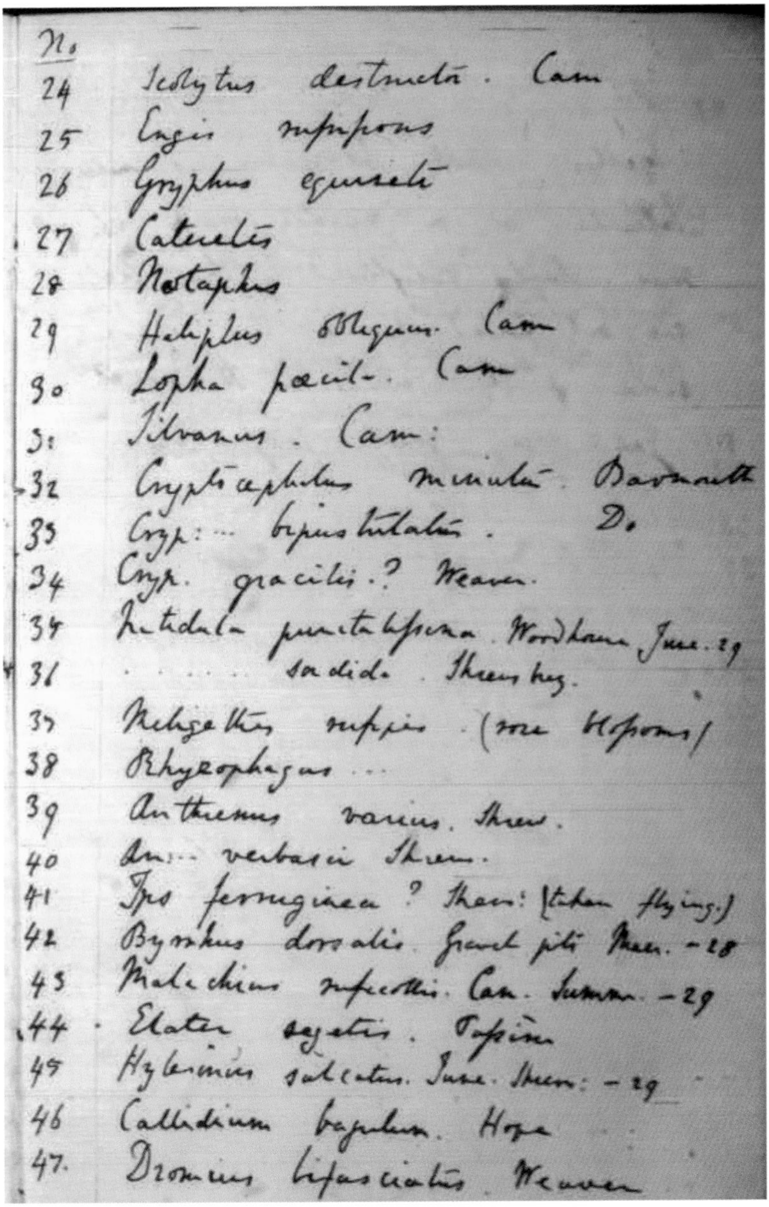

A page from Darwin's *Edinburgh notebook* (DAR 118) listing beetle specimens together with their location of collection or the name of the person who gave him a specimen. Courtesy of the Syndics of Cambridge University Library.

Cartoons of Darwin collecting beetles by Albert Way. (DAR 204:19). Courtesy of the Syndics of Cambridge University Library.

Darwin clearly impressed others with his concentrated attention and attainments in beetle collecting. He later recalled: "I remember one of my sporting friends, [James Farley] Turner, who saw me at work on my beetles, saying that I should some day be a Fellow of the Royal Society, and the notion seemed to me preposterous." (Darwin was elected a Fellow of the Royal Society in 1839.) However in these early days he was still unable to remember the Latin species names. A family friend remembered that near the end of his life Darwin told her "he found the greatest difficulty in remembering their names so that his fellow students would often for a joke to pick up the commonest kind of beetles, & put them down suddenly before him, & say 'now Darwin you're the man for beetles, what's the name of that?' — for the life of him he could not remember."[14]

Darwin left Christ's on 8 June 1829 for London. He returned to Shrewsbury and continued his usual lifestyle of shooting and beetling in Shropshire and Staffordshire. In mid-June, he went on an entomological tour in North

[14] Forster, L. M. 1883. [Recollections of Darwin, January.] (DAR 112.A31-A37), transcribed by Kees Rookmaaker. *The Complete Work of Charles Darwin Online* (http://darwin-online.org.uk/).

Wales with F. W. Hope. This trip was cut short when Darwin's recurrant dermatitus of the lips flared up again. He wrote to Fox: "the two first days I went on pretty well, taking several good insects, but for the rest of that week, my lips became suddenly so bad, & I myself not very well, that I was unable to leave the room, & on the Monday I retreated with grief & sorrow back again to Shrewsbury".[15] Although Darwin had further plans to enlarge his beetle collection and shoot throughout the summer, he also needed to prepare for the Previous Examination. "I must read for my little Go. Graham smiled & bowed so very civilly, when he told me that he was one of the six appointed to make the examination stricter, & that they were determined they would make it a very different thing from any previous examination that from all this, I am sure, it will be the very devil to pay amongst all idle men & Entomologists."[16] He then spent another week at Maer for the usual shooting and visiting. In early October Darwin attended a music meeting in Birmingham with his Wedgwood cousins.

He returned to Christ's on 12 October 1829. He entomologized with the clergyman naturalist Leonard Jenyns (1800–1893), vicar of nearby Swaffham Bulbeck. Jenyns (later named Blomefield) recalled many years later:

> [Darwin] was at that time a most zealous Entomologist...He occasionally visited me at my Vicarage, at Swaffham Bulbeck, and we made Entomological excursions together, sometimes in the Fens — that rich district yielding so many rare species of insects and plants — at other times in the woods and plantations of Bottisham Hall. He mostly used a sweeping net, with which he made a number of successful captures I had never made myself, though a constant resident in the neighbourhood.[17]

At first Darwin had seen Jenyns as a collecting competitor, recalling in later years: "At first I disliked him from his somewhat grim and sarcastic expression; and it is not often that a first impression is lost; but I was completely mistaken and found him very kind-hearted, pleasant and with a good stock of humour. I visited him at his parsonage on the borders of the Fens [Swaffham Bulbeck], and

[15] To W. D. Fox [3 July 1829], *Correspondence*, vol. 1: 87.

[16] To W. D. Fox [15 July 1829], *Correspondence*, vol. 1: 89.

[17] Blomefield, L. 1887. *Chapters in my life*. Bath: [privately printed], p. 44.

had many a good walk and talk with him about Natural History."[18] After the voyage of the *Beagle*, Jenyns would describe Darwin's fish specimens.[19]

Darwin became a fixture at the Friday evening soirées at the home of the Reverend John Stevens Henslow (1796–1861), Professor of Botany and Fellow of St John's, where the scientifically-minded of the University, young and old, were welcomed.[20] Henslow, the former professor of Mineralogy, was only 32 and for the past three years Professor of Botany. Darwin's eventual friendship with Henslow "influenced my whole career more than any other". Henslow lived at Gothic Cottage on Regent Street, which no longer survives. Here Darwin was able to listen in on the conversation of the great philosopher polymath William Whewell (1794–1866), the new professor of mineralogy. Darwin thought him "Next to Sir J. Mackintosh he was the best converser on grave subjects to whom I ever listened." and the mathematician astronomer George Peacock (1791–1858) and perhaps the great astronomer John Herschel (1792–1871) from London. Jenyns, Henslow's brother-in-law, was also a regular. As Darwin later wrote in a memorial to the late Henslow:

> Once every week he kept open house in the evening, and all who cared for natural history attended these parties, which, by thus favouring intercommunication, did the same good in Cambridge, in a very pleasant manner, as the Scientific Societies do in London. At these parties many of the most distinguished members of the University occasionally attended; and when only a few were present, I have listened to the great men of those days, conversing on all sorts of subjects, with the most varied and brilliant powers. This was no small advantage to some of the younger men, as it stimulated their mental activity and ambition.[21]

---

[18] *Autobiography*, pp. 66–67.

[19] Jenyns, Leonard.1842. *Fish.* Part 4 of *The zoology of the voyage of HMS Beagle.* Edited and superintended by Charles Darwin. London: Smith Elder and Co.. http://darwin-online.org.uk/content/frameset?viewtype=text&itemID=F9.4&pageseq=1.

[20] On Henslow, see Walters, S.M. and Stow, E.A. 2001. *Darwin's Mentor: John Stevens Henslow, 1796–1861.* Cambridge: Cambridge University Press; Barlow, Nora ed. 1967. *Darwin and Henslow. The growth of an idea.* London: Bentham-Moxon Trust, John Murray; and Darwin, C. R. 1862. [Recollections of Professor Henslow]. In Jenyns, L., *Memoir of the Rev. John Stevens Henslow M.A., F.L.S., F.G.S., F.C.P.S., late Rector of Hitcham and Professor of Botany in the University of Cambridge.* London: John Van Voorst, pp. 51–55 on *Darwin Online.* See also Darwin's reminiscences of Henslow in his *Autobiography.*

[21] Jenyns, Leonard, 1862. *Memoir of the Rev. John Stevens Henslow M.A., F.L.S., F.G.S., F.C.P.S., late Rector of Hitcham and Professor of Botany in the University of Cambridge*, p. 52.

John Stevens Henslow. Lithograph by Thomas Herbert Maguire, 1849.

Lithograph of William Whewell, 1835.

Darwin also enrolled in Henslow's university botany lectures from 1829 to 1831. Darwin recalled that he "liked them much for their extreme clearness, and the admirable illustrations; but I did not study botany. Henslow used to take his pupils, including several of the older members of the University, field excursions, on foot, or in coaches to distant places, or in a barge down the river, and lectured on the rarer plants or animals which were observed. These excursions were delightful."[22] Many years later Darwin referred to his friendship with Henslow as "a circumstance which influenced my whole career more than any other".

[22] *Autobiography*, p. 60.

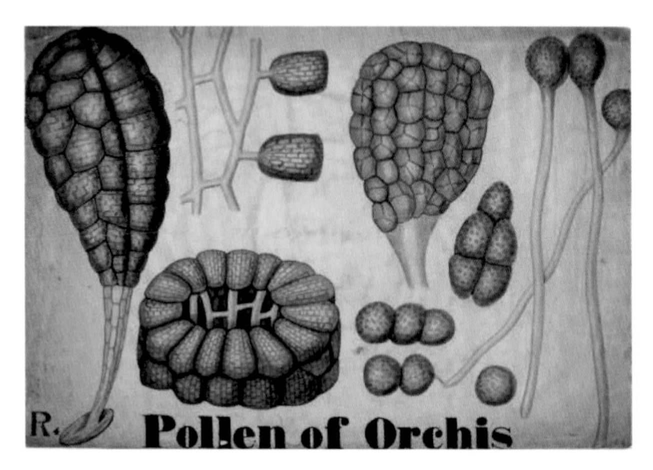

One of the "admirable illustrations" used in Henslow's botany lectures. Reproduced courtesy of the Whipple Museum of the History of Science, University of Cambridge. Darwin's study of orchids was the first book he published after *Origin of Species*. He demonstrated that natural selection could explain the wonderful adaptations of the orchid's pollen packets or pollinia.[23]

> ...during the latter half of my time at Cambridge [I] took long walks with [Henslow] on most days; so that I was called by some of the dons "the man who walks with Henslow"; and in the evening I was very often asked to join his family dinner. His knowledge was great in botany, entomology, chemistry, mineralogy, and geology. His strongest taste was to draw conclusions from long-continued minute observations.[24]

This latter trait, surely, rubbed off on Henslow's young pupil and admirer.

A few years later, in April 1832, when he encountered tropical rainforest near Rio de Janeiro at the beginning of the voyage of the *Beagle*, Darwin was reminded of an engraving he had studied and admired at Henslow's house. Darwin wrote to Henslow from Brazil: "I first saw a Tropical forest in all its sublime grandeur. — Nothing, but the reality can give any idea, how wonderful, how magnificent the scene is. — If I was to specify any one thing I should give the preemenence to the host of parasitical plants. — Your engraving is exactly true, but underates, rather

[23] Darwin, C.R. 1862. *On the various contrivances by which British and foreign orchids are fertilised by insects.* London: John Murray.

[24] *Autobiography*, p. 64.

than exagérâtes the luxuriance. — I never experienced such intense delight."[25] At the time Darwin wrote in his *Beagle diary*:

> After passing through some cultivated country we entered a Forest, which in the grandeur of all its parts could not be exceeded. — As the gleams of sunshine penetrate the entangled mass, I was forcibly reminded of the two French engravings after the drawings of Maurice Rugendas & Le Compte de Clavac. — In these is well represented the infinite numbers of lianas & parasitical plants & the contrast of the flourishing trees with the dead & rotten trunks. I was at an utter loss how sufficiently to admire this scene.[26]

A scene that made Darwin long for the tropics. "La Forêt du Brésil", engraving by C.O.F. Jean-Baptiste, Comte de Clarac, 1819.

To give an example of Henslow's noble character, Darwin mentioned some of his own scientific work that might otherwise have gone unrecorded:

> Whilst examining some pollen-grains [under a microscope] on a damp surface I saw the tubes exserted, and instantly rushed off to communicate my surprising discovery to him. Now I do not suppose any other Professor of Botany could

[25] Darwin to J. S. Henslow 18 May–16 June 1832, *Correspondence*, vol. 1: 236.

[26] *Beagle diary*, p. 53.

> have helped laughing at my coming in such a hurry to make such a communication. But he agreed how interesting the phenomenon was, and explained its meaning, but made me clearly understand how well it was known; so I left him not in the least mortified, but well pleased at having discovered for myself so remarkable a fact, but determined not to be in such a hurry again to communicate my discoveries.[27]

Another episode experienced with Henslow captured a dramatic scene in the streets of Cambridge:

> I once saw in his company in the streets of Cambridge almost as horrid a scene, as could have been witnessed during the French Revolution. Two body-snatchers had been arrested and whilst being taken to prison had been torn from the constable by a crowd of the roughest men, who dragged them by their legs along the muddy and stony road. They were covered from head to foot with mud and their faces were bleeding either from having been kicked or from the stones; they looked like corpses, but the crowd was so dense that I got only a few momentary glimpses of the wretched creatures. Never in my life have I seen such wrath painted on a man's face, as was shown by Henslow at this horrid scene. He tried repeatedly to penetrate the mob; but it was simply impossible. He then rushed away to the mayor, telling me not to follow him, to get more policemen. I forget the issue, except that the two were got into the prison before being killed.[28]

In early November, his brother Erasmus stayed with Darwin for a few days. The brothers spent long hours at the gallery of the Fitzwilliam Museum. At the end of Darwin's second year at Christ's College, he had settled into a agreeable world of genteel pleasures and pastimes mixed with wide reading and the access to a wider world of learning that universities can still offer to those who seek it. There were signs that he might have some promise as a mature naturalist, but nothing that would suggest the path his career would ultimately take.

---

[27] *Autobiography*, p. 66.

[28] *Autobiography*, p. 65.

THIRD YEAR AT CAMBRIDGE

After the Christmas break, Darwin returned from London to Christ's on New Year's Day 1830. About this time, Darwin and seven friends formed an informal weekly dining club called either the "Gourmet Club" or "Glutton Club". The club consisted of Darwin, Herbert, Whitley, Watkins, Cameron, James Heaviside of Sidney Sussex College, Robert Blane of Trinity College and Henry Lowe of Trinity Hall. They dined in each other's rooms in rotation. Herbert later recalled the formation of the club in a letter to Darwin's son Francis Darwin:

> At our first meeting Cameron proposed that we should call ourselves the "Glutton Club" to show our contempt for another set of men who called themselves by a long Greek title, meaning — "fond of dainties;" but who falsified their claim to such a designation by their weekly practice of dining at some road-side inn, 6 miles from Cambridge, on Mutton Chops, or Beans & Bacon. The name we adopted was in truth most dysphemistic, if I may use such a word; for we were none of us given to excess, but our little dinners were very recherchés & well-served; & we generally wound up the evening with a game of mild Vingt-&-un.[98]

[98] Herbert, J. M. 1882. [Recollections of Darwin, 12 June.] (DAR 112.A60-A61), transcribed by Kees Rookmaaker. *The Complete Work of Charles Darwin Online* (http://darwin-online.org.uk/).

"Christ College from the street". R. Ackermann's, *History of Cambridge*, 1815.

Another member of the club, Watkins, remembered it somewhat differently:

> [Darwin] was a member of the "Gourmet Club" — so called not because its members were gluttons, but because they made a devouring raid on birds & beasts which were before unknown to human palate. Our menu was certainly a choice one but the appetite for strange flesh did not last very long & I think the Club came to an untimely end by endeavouring to eat an old brown owl which was indescribable! We tried hawk & bittern & other delicacies wh. I have forgotten.[99]

Cambridge did not consist exclusively of dining, shooting and beetling for Darwin. In addition to his College lectures, readings and examinations he had to pass the two set University examinations. On Wednesday 24 March 1830, after weeks of fervent cramming, Darwin nervously entered the neoclassical Senate House to take the Previous Examination or 'Little Go'. Candidates were examined orally, in turn. Three hours in the morning were spent on the classics and three hours

[99] Watkins, F. [1887]. [Recollections of Darwin, 18 July.] (DAR 112.A111-A114), transcribed by Kees Rookmaaker, *The Complete Work of Charles Darwin Online* (http://darwin-online.org.uk/).

in the afternoon on the New Testament and Paley. The results were posted the following day, when Darwin wrote to Fox triumphantly:

My dear Fox

> I am through my little Go,!!! am too much exalted to humble myself by apologising for not having written before. — But, I assure you before I went in & when my nerves were in a shattered & weak condition, your injured person often rose before my eyes & taunted me with my idleness. But I am through through through. I could write the whole sheet full, with this delightful word. — I went in yesterday, & have just heard the joyful news. — I shall not know for a week, which class I am in. — The whole examination is carried on in a different system. It has one grand advantage, being over in one day. They are rather strict; & ask a wonderful number of questions[100]

Darwin's Cambridge Beetle Cabinets

In March, Darwin wrote Fox "My new [beetle] cabinet is come down & a gay little affair it is." In a later letter written from Shrewsbury, Darwin described this cabinet from memory: "The man who made my cabinet is W. Edwards 29 Wilton St. Westminster. I advise to get one bigger than mine. — Mine cost £5.10. & contained 6 drawers, depth, $1^{f.}$3. breadth $1^{f.}$7. — & whole cabinet stood in height 1"4."[101] In April, Darwin wrote to Fox: "I find I get on very slowly with my cabinet, & shall be very glad of your assistance. I have only yet got to the Amarœ. — … I have been seeing a good deal lately of Prof. Henslow; I took a long walk with him the other day: I like him most exceedingly, he is so very goodnatured & agreeable".[102]

In 2007 I was visiting my late friend Dr Milo Keynes, a great-grandson of Darwin. He still had vivid memories of his uncle Lenny, Darwin's fourth son Leonard (1850–1943). Milo's house on Midsummer Common was crammed with beautiful engravings and books and a few Darwin relics. There was a dining room chair that Darwin bought when he married Emma in 1839. Next was a tiny onyx cross and plinth that belonged to Emma and Darwin's soapstone Chinese letter seal. There were also some tables said to come from Down House. Of particular interest was an exquisite colour miniature of Darwin's mother,

[100] *Correspondence*, vol. 1: 101. See Pass Lists, Previous Examinations 1824–1883, Exam. L. 9, Cambridge University Archives.

[101] *Correspondence*, vol. 1: 127.

[102] *Correspondence*, vol. 1: 102.

Susanah Wedgwood, painted by Peter Paillou in 1793, before her marriage. She looked just like Darwin, without the beard of course. On the back side was a braided lock of brown hair under glass.

Milo then pointed to a small wooden cabinet on which his small television perched. He said it too had belonged to Darwin and had come from Down House. Milo thought it might have been used on the *Beagle* as it was a rather plain and unadorned dark mahogany box with a brass lock and key.

I opened it and was astonished to see that it was an insect collecting cabinet with six drawers — the same number that Darwin mentioned in his letter to Fox. Milo recalled that the cabinet contained shells when he inherited it from his mother. Its measurements resemble the one described by Darwin. I later sent photographs to the firm of Cheffins Antiques in Cambridge. They estimated the cabinet dates from the early nineteenth century.

The drawers have cork-lined bases for receiving insect pins. Three have a camphor compartment at the front to prevent an insect collection from being damaged by pests. At some later point, the bottoms of the drawers were lined with paper. In some places the paper is missing, revealing the cork. In one of these exposed areas in the third drawer I spotted two very fine insect pins stuck in the cork, and bent over flat. Inside one of the camphor compartments was a minute printed paper label: "Ligustri Privet.", (the privet hawk moth). This label almost certainly dates from after Darwin's time.

The mahogany insect cabinet which belonged to Dr Milo Keynes.

The cabinet on display in Darwin's restored College rooms in 2009. Photo Allison Maletz.

The dimensions in Darwin's letter are: 19 wide, 16 high, 15 deep. Milo's cabinet measures: 16 wide, 19 high, 19 deep. Although not identical, both measurements refer to a very deep rectangular box. Milo kindly agreed to lend his cabinet to Christ's College to be displayed in Darwin's rooms when they were opened to the public in 2009. If it *was* one of Darwin's Cambridge beetle cabinets, then it returned home after 178 years. After the rooms were restored, as discussed below, the cabinet was the crowning glory of the display. After the bicentennary year, and the cabinet was returned, an excellent replica cabinet was constructed by Martin Tuck of the College Maintenance Department.

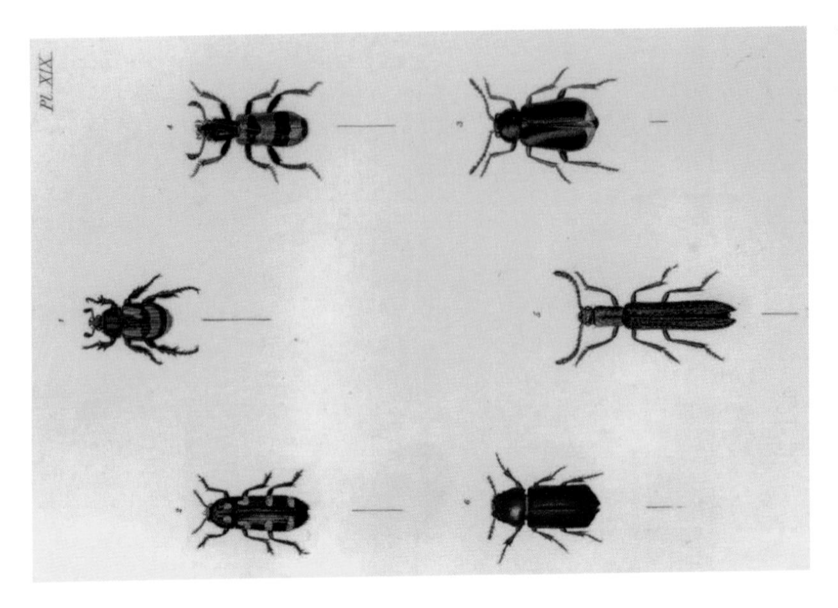

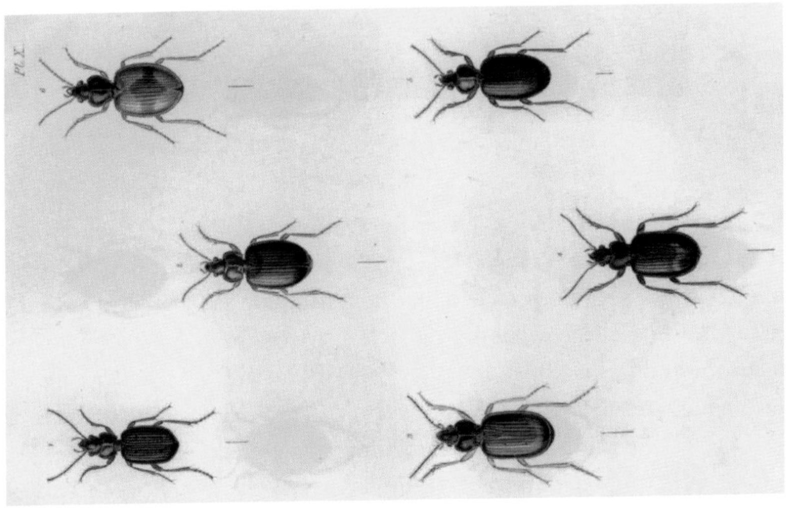

Two plates of beetles that Darwin would have seen while a student at Cambridge. Stephens, *Illustrations of British entomology*, Mandibulata, (1829–1832) vols. 2 & 3.

The New Bridge. St John's. 1840. T. D. Atkinson. *Cambridge described and illustrated.* Macmillan, 1897.

Another Term

On 3 June 1830, Darwin was recorded as leaving Christ's at the end of Easter Term. He spent a few days in London before heading north to Shrewsbury. In August, he returned to North Wales for beetle collecting and fishing followed by the usual rounds of shooting in Shropshire and Staffordshire. In September, he wrote to Fox about his new horse which he hoped would make a very good hunter. Darwin took the horse with him as he trotted back amongst the medieval spires and narrow winding lanes of Cambridge on 7 October 1830, for the start of the Michaelmas Term. He wrote to Fox: "I arrived here in my most snug & comfortable rooms yesterday evening". Herbert recalled an incident which occurred at the beginning of this term.

> Darwin asked me to take a long walk with him in the Fens, to search for some natural objects he was desirous of having. After a very fatiguing day's work we dined together late in the evening, at his rooms in X[ts]. Coll:, and as soon as our dinner was over, we threw ourselves into easy chairs, & fell sound asleep. I was the first to awaken, about 3 in the morning, when having looked at my watch, & knowing the strict rule of St. John's, which required men in statu pupillari to come into college before midnight, I rushed homeward at the utmost speed, in fear of the consequences, but hoping that the Dean w[d] accept the excuse as sufficient, — when I told him the real facts. He, however, was inexorable, and refused to receive my explanations, or any evidence I c[d]. bring; and although during my undergraduateship I had never been reported for coming late into College — now, when I was a hard-working B.A., and had 5 or 6 pupils, he sentenced me to confinement to the College Walls for the rest of the term."[103]

Darwin was outraged by Herbert's punishment, but then he was accustomed to the relaxed atmosphere at Christ's. This recollection also reveals that Darwin had two easy chairs in his rooms. In November, Darwin wrote to Fox that because of all the reading involved in "getting up all my subjects" he was left no time to catch or send insects.

Darwin paid Henslow to act as his private tutor in mathematics. Darwin wrote to Fox that "the hour with him is the pleasantest in the whole day".

---

[103] Herbert, J. M. 1882. [Recollections of Darwin, 12 June.] (DAR 112.A60-A61), transcribed by Kees Rookmaaker. *The Complete Work of Charles Darwin Online* (http://darwin-online.org.uk/).

At Henslow's botany lectures Darwin helped to arrange the specimens and materials before the lectures, and was generally considered to be Henslow's favourite. William Allport Leighton (1805–1889), an old Shrewsbury school friend, recalled: "I remember that the Professor in the concluding remarks at the close of his course of lectures said he hoped his teaching had influenced many to perseverance — certainly he knew it had influenced one — no doubt he meant Darwin."[104]

Darwin spent the Christmas vacation of 1830 in Cambridge preparing for his final examination. He later recalled this time in his *Autobiography*:

> in my last year I worked with some earnestness for my final degree of B.A., and brushed up my Classics together with a little Algebra and Euclid, which latter gave me much pleasure, as it did whilst at school. In order to pass the B.A. examination, it was, also, necessary to get up Paley's *Evidences of Christianity*, and his *Moral Philosophy*. This was done in a thorough manner, and I am convinced that I could have written out the whole of the Evidences with perfect correctness, but not of course in the clear language of Paley. The logic of this book and as I may add of his *Natural Theology* gave me as much delight as did Euclid. The careful study of these works, without attempting to learn any part by rote, was the only part of the Academical Course which, as I then felt and as I still believe, was of the least use to me in the education of my mind. I did not at that time trouble myself about Paley's premises; and taking these on trust I was charmed and convinced by the long line of argumentation.[105]

Paley famously began his book *Natural Theology* with the following analogy:

> In crossing a heath, suppose I pitched my foot against a *stone*, and were asked how the stone came to be there; I might possibly answer, that, for anything I knew to the contrary, it had lain there forever... But suppose I had found a *watch* upon the ground...There must have existed, at some time, and at some place or other, an artificer or artificers, who formed [the watch] for the purpose which we find it actually to answer; who comprehended its construction, and

[104] Leighton, W. A. c. 1886. [Recollections of Charles Darwin]. (DAR 112.B94–B98), transcribed by Kees Rookmaaker. *The Complete Work of Charles Darwin Online* (http://darwin-online.org.uk/).

[105] *Autobiography*, pp. 58–59. See Fyfe, A. 1997. The reception of William Paley's *Natural Theology* in the University of Cambridge. *British Journal for the History of Science* 30: 321–335 which corrected the error in some accounts that Darwin was required to read Paley's *Natural Theology*.

designed its use. ... every indication of contrivance, every manifestation of design, which existed in the watch, exists in the works of nature...[106]

Hence, Paley's argument was built first on the premise that nature reveals purpose and design. Secondly, purpose and design must be the product of a designer. And thus he concluded: nature is the product of a designer. The designer, for the Rev. Paley, was of course the God of the Church of England.[107] Nevertheless by bringing together the clockwork *Weltanschauung* of Descartes' mechanical universe and the support of evidence from the natural world to justify religious faith, he may have inadvertently opened the way for some of his followers, such as the young Darwin, to eventually doubt and reject that faith if the premises and evidence of nature pointed to other conclusions.

Only after the voyage of the *Beagle* would Darwin reject Paley's argument of design, already quite outdated by the 1830s. Cambridge continued to make use of Paley for many years. Hanging as a framed curiosity in Christ's Old Library is the last University Previous Examination paper on Paley's *Evidences of Christianity* from 9 December 1920.

Darwin emphasized how his own work superseded Paley's argument in his *Autobiography*:

> The old argument of design in nature, as given by Paley, which formerly seemed to me so conclusive, fails, now that the law of natural selection has been discovered. We can no longer argue that, for instance, the beautiful hinge of a bivalve shell must have been made by an intelligent being, like the hinge of a door by man. There seems to be no more design in the variability of organic beings and in the action of natural selection, than in the course which the wind blows. Everything in nature is the result of fixed laws. But I have discussed this subject at the end of my book on the *Variation of Domestic Animals and Plants*, and the argument there given has never, as far as I can see, been answered.[108]

---

[106] Paley, W. 1802. *Natural theology: or, evidences of the existence and atributes of the deity, collected from the appearances of nature.* London, pp. 1–4, 19.

[107] I have adopted the summary of LeMahieu, D.L. 1976. *The Mind of William Paley: a philosopher and his age.* Lincoln, Nebraska, p. 58.

[108] *Autobiography*, pp. 87–88.

As many readers may be unfamiliar with Darwin's book on *Variation*, the passage to which he refers is given below.

> Some authors have declared that natural selection explains nothing, unless the precise cause of each slight individual difference be made clear. Now, if it were explained to a savage utterly ignorant of the art of building, how the edifice had been raised stone upon stone, and why wedge-formed fragments were used for the arches, flat stones for the roof, &c.; and if the use of each part and of the whole building were pointed out, it would be unreasonable if he declared that nothing had been made clear to him, because the precise cause of the shape of each fragment could not be given. But this is a nearly parallel case with the objection that selection explains nothing, because we know not the cause of each individual difference in the structure of each being.
>
> The shape of the fragments of stone at the base of our precipice may be called accidental, but this is not strictly correct; for the shape of each depends on a long sequence of events, all obeying natural laws; on the nature of the rock, on the lines of deposition or cleavage, on the form of the mountain which depends on its upheaval and subsequent denudation, and lastly on the storm or earthquake which threw down the fragments. But in regard to the use to which the fragments may be put, their shape may be strictly said to be accidental. And here we are led to face a great difficulty, in alluding to which I am aware that I am travelling beyond my proper province. An omniscient Creator must have foreseen every consequence which results from the laws imposed by Him. But can it be reasonably maintained that the Creator intentionally ordered, if we use the words in any ordinary sense, that certain fragments of rock should assume certain shapes so that the builder might erect his edifice? If the various laws which have determined the shape of each fragment were not predetermined for the builder's sake, can it with any greater probability be maintained that He specially ordained for the sake of the breeder each of the innumerable variations in our domestic animals and plants; — many of these variations being of no service to man, and not beneficial, far more often injurious, to the creatures themselves? Did He ordain that the crop and tail-feathers of the pigeon should vary in order that the fancier might make his grotesque pouter and fantail breeds? Did He cause the frame and mental qualities of the dog to vary in order that a breed might be formed of indomitable ferocity, with jaws fitted to pin down the bull for man's brutal sport? But if we give up the principle in one case, — if we do not admit that the variations of the primeval dog were intentionally guided in order that the greyhound, for instance, that perfect image of symmetry and vigour, might be formed, — no shadow of reason can

be assigned for the belief that variations, alike in nature and the result of the same general laws, which have been the groundwork through natural selection of the formation of the most perfectly adapted animals in the world, man included, were intentionally and specially guided. However much we may wish it, we can hardly follow Professor Asa Gray in his belief "that variation has been led along certain beneficial lines," like a stream "along definite and useful lines of irrigation." If we assume that each particular variation was from the beginning of all time preordained, the plasticity of organisation, which leads to many injurious deviations of structure, as well as that redundant power of reproduction which inevitably leads to a struggle for existence, and, as a consequence, to the natural selection or survival of the fittest, must appear to us superfluous laws of nature. On the other hand, an omnipotent and omniscient Creator ordains everything and foresees everything. Thus we are brought face to face with a difficulty as insoluble as is that of free will and predestination.[109]

[109] Darwin, C. R. 1868. *The variation of animals and plants under domestication.* London: John Murray. 1st ed, 1st issue. vol. 2, pp. 430–432.

LAST TERMS AT CAMBRIDGE

Darwin's final examination for the B.A. degree took place between 14–20 January 1831. It consisted of three days of written papers. Darwin was a poll candidate. The poll (an abbreviation of the Greek 'Hoi Polloi' for the crowd) consisted of those students who took an ordinary pass degree rather than an honours degree. The examination consisted of six parts: Homer, Virgil, Euclid, arithmetic and algebra, Paley's *Evidences of Christianity* (1794) and *Principles of moral and political philosophy* (1764), and Locke's *An essay concerning human understanding* (1690).

The papers were marked by Friday 21 January. On Saturday the 22nd, a second edition of the *Cambridge Chronicle* appeared which showed Darwin placed 10 out of 178 in the polls. There were only 86 honours candidates that year. The following day Darwin wrote to Fox: "I sent you a newspaper yesterday, in which you will see what a good place I got in the Polls". As Darwin recalled in his *Autobiography*, p. 59, "By answering well the examination questions in Paley, by doing Euclid well, and by not failing miserably in Classics, I gained a good place among the ὁι πολλοι or crowd of men who do not go in for honours". However, as Darwin had not resided the requisite number of terms he could not yet be awarded his degree. His final two terms at Christ's, without the pressure of preparing for examinations, were some of the most pleasant and important he spent in Cambridge.

156

THE METHOD OF PROCEEDING TO THE DEGREE OF BACHELOR OF ARTS.

STUDIES AND EXERCISES.

The ordinary course of study preparatory to the degree of B.A. may be considered under the three heads of Natural Philosophy, Theology and Moral Philosophy, and the Belles Lettres. On these subjects, independently of the *public* lectures which are delivered by the several Professors in the University, the Students attend the lectures of the Tutors of their respective colleges; and the instructions comprehended in the three general heads above-named may be thus stated. In the *first*, Euclid's Elements, the Principles of Algebra, Plane and Spherical Trigonometry, Conic Sections, Mechanics, Hydrostatics, Optics, Astronomy, Fluxions, Newton's Principia, Increments, &c. &c. In the *second*, Beausobre's Introduction, Doddridge's and Paley's Evidences, the Greek Testament, Butler's Analogy, Paley's Moral Philosophy, Locke's Essay, and Duncan's Logic. In the *third*, the most celebrated Greek and Latin Classics.

Besides a constant attendance on college lectures, the Undergraduates are examined yearly, or half-yearly, in those subjects which have engaged their studies, and according to the manner in which they acquit themselves in these examinations, their names are arranged in classes, and those who obtain the honour of a place in the first class receive, according to merit, prizes of books, of different value.

By this mode of procedure the Students are prepared for those *public* examinations and exercises which the University requires of all candidates for degrees. The first of these examinations takes place in the *fifth* term of academical residence, according to the following plan.

Regulations for the Institution of a previous *Examination of all persons who take the Degrees of B.A. B.C.L. or M.B.—(Confirmed by Graces of the Senate, March* 13, 1822, *and December* 7, 1825.*)*

The *University Calendar* prescribed the curriculum for the B.A. degree.

In February Darwin read the astronomer John Herschel's *Preliminary discourse on the study of natural philosophy* (1831), an authoritative and thought-provoking model on correct methods of scientific investigation. It was essentially a survey of the science of the day, as well as an account of the progress science had so far achieved. Herschel's law of continuity meant that all parts of nature and science would be interconsistent. Given the collection of enough facts, powerful

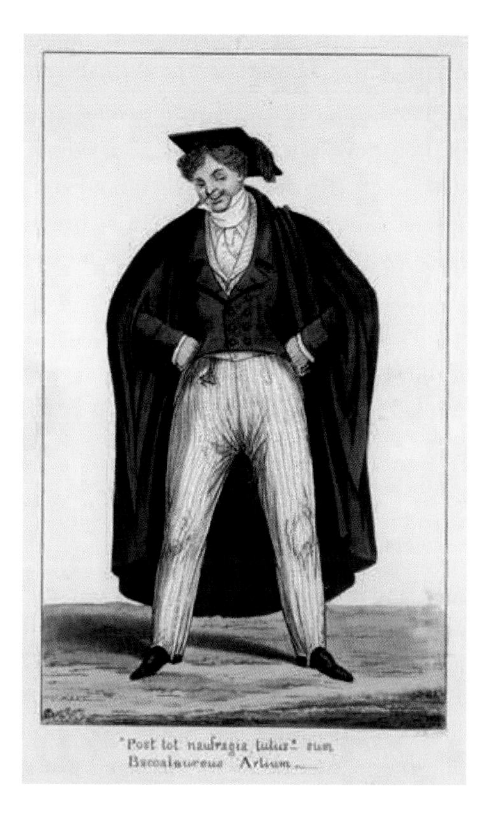
"Post tot naufragia tutus sum
Baccalaureus Artium

"Safe after such a shipwreck I am bachelor of Arts". *Gradus ad Cantabrigiam; or, the new university guide* (1824).

general laws could be deduced. Darwin also read Alexander von Humboldt's *Personal narrative* of his expedition to northern parts of South America.

> I read with care and profound interest Humboldt's Personal Narrative. This work and Sir J. Herschel's Introduction to the Study of Natural Philosophy stirred up in me a burning zeal to add even the most humble contribution to the noble structure of Natural Science. No one or a dozen other books influenced me nearly so much as these two. I copied out from Humboldt long passages about Teneriffe, and read them aloud on one of [Henslow's] excursions[110]

[110] *Autobiography*, pp. 68–69. See the important essay by Gordon Chancellor, Humboldt's *Personal narrative* and its influence on Darwin http://darwin-online.org.uk/EditorialIntroductions/Chancellor_Humboldt.html.

Some years later Darwin described his time in Cambridge after his degree: "During these months lived much with Prof. Henslow, often dining with him, & walking with. became slightly acquainted with several of the learned men in Cambridge. which much quickened the little zeal, which dinner parties & hunting had not destroyed. In the Spring...talked over an excursion to Teneriffe."[111] Fox heard some of Darwin's almost breathless enthusiasm: "At present, I talk, think, & dream of a scheme I have almost hatched of going to the Canary Islands. — I have long had a wish of seeing Tropical scenery & vegetation: & according to Humboldt Teneriffe is a very pretty specimen."[112] He also started learning Spanish, which would stand him in good stead during the voyage of the *Beagle*. The planned trip to Tenerife, in the footsteps of Humboldt, never materialized.

John Medows Rodwell (1808–1900) accompanied him on some of Henslow's excursions and recalled some of Darwin's exploits:

> In one of Professor Henslows Botanizing excursions to Bottisham Fen I well recollect an amusing incident wh[ich] befel Darwin. In order to clear the ditches we were provided with several jumping poles with which we had to swing ourselves across. One object of our search was to find the [Atriculavia], a specimen of which caught his keen eye, and in order to secure he attempted to jump the ditch on the opposite side of which it grew. Not however having secured sufficient impetus for the leap, the pole stuck fast in the middle in a vertical position, of course with D[arwin] at the top. Nothing daunted however he coolly slid down, secured the prize, and brought it, all much besmirched as he was to the amused Professor.
>
> We once had a very amusing expedition to Gamlingay heath in search of Natter-jacks [a toad]. Darwin was very succesful in detecting the haunts of these pretty reptiles and catching them. He brought several to Profr Henslow who said laughingly — well Darwin "are you going to make a Natter-jack pie?" It was on this day that he was very succesful in finding plants: particularly the Anemone pulsatillus and the Colchecum antennuatus which had never before been found except on the Gogmagogs and in L'Osbornes park.
>
> My acquaintance with Mr Darwin was made at Professor Henslow's House, at whose soirees and lectures he was always present and was most useful to the professor in arranging specimens and getting the room in order for Lectures. It

---

[111] *Journal*, p. 8.

[112] To W. D. Fox [7 April 1831], *Correspondence*, vol. 1: 120.

> was obvious that Darwin was Henslow's favourite pupil & that he saw in him of prognostications of future distinction & eminence as a naturalist. One feature of D[arwin'].s mind I particularly used to notice — viz. his determination to prosecute all his investigations to the very bottom. Professor Henslow used to say "What a fellow that D[arwin]. is for asking questions!" I happened one day to mention some rather rare plant growing in the neighbourhood of Bury St. Edmunds, which I promised to send him — but forgot to do so — & soon received a very amusing letter to say that if I did not send them he would come for them in propria persona & charge me for the journey![113]

Darwin later recorded in 1838, "In the Spring [1831], Henslow persuaded me to think of Geology & introduced me to Sedgwick."[114] However Darwin wrote in his *Autobiography* in 1876 that "Public lectures on several branches were given in the University, attendance being quite voluntary; but I was so sickened with lectures at Edinburgh that I did not even attend Sedgwick's eloquent and interesting lectures." Yet, the recollections of some university friends give such specific details (and Darwin himself noted that they were eloquent and interesting) that it seems he did attend Sedgwick's lectures at some point. Darwin might have meant that he did not attend any before his degree. Rodwell remembered an occasion when Darwin not only impressed Sedgwick, but started an undergraduate craze:

> Professor Sedgwick happened one day to mention a spring issuing from one of the chalk hills at Trumpington or Coton which deposited carb[onate]. of lime very prettily upon twigs &c. Darwin said to me, "I shall go and test that water for myself", which he did and found the fact to be as Sedgwick had stated it. Not content with this he deposited a large bush in the spring and at a subsequent lecture presented it to Sedgwick who exhibited it as being, what it really was, a very beautiful specimen. Several members of Sedgwick's class followed D[arwin]'s example and adorned their rooms with similar specimens of Increstation.[115]

---

[113] Rodwell, J. M. 1882. [Recollections of Darwin, 8 July.] (DAR 112.A94-A95), transcribed by Kees Rookmaaker. *The Complete Work of Charles Darwin Online* (http://darwin-online.org.uk/).]

[114] *Journal*, p. 8.

[115] Rodwell, J. M. nd. [Recollections of Darwin in Cambridge.] (DAR112.B118-B121) , transcribed by Kees Rookmaaker. *The Complete Work of Charles Darwin Online* (http://darwin-online.org.uk/).

Elsewhere Rodwell recalled a geological conversation with Darwin:

> we were talking over one of Sedgwick lectures in wh. he had spoken of the enlarged views both of Time & Space what Geology could give. he said to me — It strikes me that all our knowledge about the structure of our Earth is very much like what an old hen c[d] know of the hundred-acre field in a corner of which she is scratching! — & afterwards, "what a capital hand is Sedgwick for drawing large cheques upon the Bank of Time!" — which of course was in reference to some speculation of Sedgwick's as to the probable antiquity of the world.[116]

In an April 1831 letter to his sister Caroline, Darwin made a rare reference to politics, but only to show his lack of interest: "The Election here is a great bore, as Henslow is Lord Palmerston's right-hand man, and he has no time for walks. — All the while I am writing now my head is running about the Tropics: in the morning I go and gaze at Palm trees in the hot-house and come home and read Humboldt: my enthusiasm is so great that I cannot hardly sit still on my chair."[117] As Janet Browne noted "[Darwin's] Cambridge was an easy-going affair, for the most part happily engaged with the internal world of his college in preference to any of the wider issues that might rampage outside."[118]

Henslow had originally been a Tory but followed Lord Palmerston when he changed to the Whigs in the spring of 1828. Because of his support of Parliamentary reform in the elections of 1831, Palmerston lost his seat as M.P. for Cambridge University. A few years later, during the Borough elections of 1835, Henslow was shocked by the accusations of bribery against the Conservative agents. He offered to serve as prosecutor. This resulted in some very unpleasant personal abuse. "Not only was the cry raised of 'Henslow, common informer', whenever he appeared in the streets, but the same obnoxious words were placarded upon the walls in such large and enduring characters that even to this day (July, 1861) they are still distinctly visible in some places." On the walls of Corpus Christi the stencilled words were still legible in the 1960s.[119]

---

116 Rodwell, J. M. 1882. [Recollections of Darwin, 8 July.] (DAR 112.A94-A95), transcribed by Kees Rookmaaker. *The Complete Work of Charles Darwin Online* (http://darwin-online.org.uk/).

117 To Caroline Darwin [28 April 1831], *Correspondence*, vol. 1: 122.

118 Browne, Janet. 1995. *Charles Darwin, vol. 1: Voyaging*. London: Pimlico, p. 93.

119 Barlow, Nora ed. 1967. *Darwin and Henslow. The growth of an idea*. London: Bentham-Moxon Trust, John Murray, p. 45.

The return of Osborn and Adam to Parliament after the election. Sidney street, Cambridge 1831.

In early May Darwin received an anonymous gift. It was a Gould-type microscope, with the accompanying note:

> If Mr. Darwin will accept the accompanying Coddington's Microscope, it will give peculiar gratification to one who has long doubted whether Mr. Darwin's talents or his sincerity be the more worthy of admiration, and who hopes that the instrument may in some measure facilitate those researches which he has hitherto so fondly and so successfully prosecuted.[120]

This small instrument was a basic compound microscope which screwed into the lid of the wooden case. Darwin later learned that the gift was from Herbert. The original microscope is now on display at Darwin's home Down House.

At the end of April Darwin's degree was conferred, along with Henry Churchman Long (1809–1884) another student from Christ's. Darwin signed the Subscriptions book "Charles Robert Darwin Christ. Coll: April 26th 1831". Two days later he wrote to his sister Caroline: "I took my Degree the other day: it cost me *£15*: there is waste of money."[121] Nevertheless Darwin's degree is officially recorded in 1832. Francis Darwin explained: "he was unable to take his degree at the usual time, — the beginning of the Lent Term, 1831. In such a case a man usually took his degree before Ash-Wednesday, when he was called "Baccalaureus ad Diem Cinerum," and ranked with the B.A.'s of the year. My

[120] *Correspondence*, vol. 1: 122.

[121] *Correspondence*, vol. 1: 122.

Darwin's signature in the Subscriptions book. Books of subscriptions for degrees, Cambridge University Archives (Matric. 11). Courtesy of the Syndics of Cambridge University Library.

father's name, however, occurs in the list of Bachelors "ad Baptistam," or those admitted between Ash-Wednesday and St. John Baptist's Day (June 24th); he therefore took rank among the Bachelors of 1832."[122]

Darwin's father gave him £200 to settle his Cambridge debts before closing this part of his life. Darwin's time as a student at Christ's College came to an end when he finally left after the Easter Term on 16 June 1831. Darwin later summarized his time in Cambridge with the disapproval of hindsight and a much older man:

> my time was sadly wasted there and worse than wasted. From my passion for shooting and for hunting and when this failed, for riding across country I got into a sporting set, including some dissipated low-minded young men. We used often

[122] F. Darwin 1887, vol. 1: 163.

Engraving of the "presentation of the Senior Wrangler to the Vice Chancellor", 1842. The Senior Wrangler was the top scoring student of mathematics.

> to dine together in the evening, though these dinners often included men of a higher stamp, and we sometimes drank too much, with jolly singing and playing at cards afterwards. I know that I ought to feel ashamed of days and evenings thus spent, but as some of my friends were very pleasant and we were all in the highest spirits, I cannot help looking back to these times with much pleasure. ... Upon the whole the three years which I spent at Cambridge were the most joyful in my happy life; for I was then in excellent health, and almost always in high spirits.[123]

But his judgement was not all negative. After all, so many intelligent and highly educated men had shown interest in him, not least his mentor Henslow.

> Looking back, I infer that there must have been something in me a little superior to the common run of youths, otherwise the above-mentioned men, so much older than me and higher in academical position, would never have allowed me to associate with them. Certainly I was not aware of any such superiority, and I remember one of my sporting friends, Turner, who saw me at work on my beetles, saying that I should some day be a Fellow of the Royal Society, and the notion seemed to me preposterous.

Darwin was elected a Fellow of the Royal Society in 1839 after the voyage of the *Beagle*. At the end of his Cambridge career, Darwin was one of the most

---

[123] *Autobiography*, pp. 60, 68.

qualified young men of science in the country. He was still young and untested to be sure. Hence, it is hardly surprising that he was offered the chance to travel on the *Beagle* as naturalist. His friend Herbert ended his letter of recollections, after Darwin's death, with a heartfelt epitaph:

> It w^{d} be idle for me to speak of his vast intellectual powers — which according to the verdict of all Europe have raised him to the foremost-place in the ranks of natural Science.
>
> But I cannot end this cursory & rambling sketch without testifying — & I doubt not all his surviving college friends w$^{d.}$ concur with me — that he was the most genial, warm-hearted, generous & affectionate of friends — that his sympathies were with all that was good & true, & that he had a cordial hatred for anything false, or vile, or cruel, or mean, or dishonourable. He was not only great — but pre-eminently good, & just, & loveable — [124]

[124] Herbert, J. M. 1882. [Recollections of Darwin, 12 June.] (DAR 112.A60-A61), transcribed by Kees Rookmaaker. *The Complete Work of Charles Darwin Online* (http://darwin-online.org.uk/).

VOYAGE OF THE *BEAGLE* — AND RETURN TO CAMBRIDGE

In August 1831 Darwin took an important and instructive geological tour of North Wales, partly in the company of Professor Adam Sedgwick.[125] Darwin became more than proficient with the surveying, measuring and hammering of practical geological field work. He learned how to read the landscape and to determine rock types, their angles of bedding and so forth. It was all part of reconstructing the past from the fragmentary evidence under foot. For the next few years, geology would move uppermost in Darwin's scientific interests. Upon returning home to Shrewsbury, Darwin received a letter from Henslow which changed his life, and the world, forever. It was an offer to travel on board the Royal Navy surveying ship HMS *Beagle* as naturalist. The offer came originally from the ship's captain, Robert FitzRoy (1805–1865) who was preparing to sail

[125] Barrett, Paul H. 1974. The Sedgwick-Darwin Geologic Tour of North Wales. *Proceedings of the American Philosophical Society* 118: 146–164; Clark, J. W. and Hughes, T.M. (eds.) 1890. The walking tour in North Wales. In *The life and letters of the Reverend Adam Sedgwick*, vol. 1: 379–381. See also Herbert, Sandra. 2002. Charles Darwin's notes on his 1831 geological map of Shrewsbury. *Archives of natural history* 29 (1): 27–30, and Lowe, Robert. 'Journal kept by H. P. Lowe & R Lowe during 3 months of the summer 1831. at Barmouth. North Wales. Forsitan haec olim meminisse juvabit.' Transcribed from the manuscript by Peter Lucas. http://darwin-online.org.uk/content/frameset?viewtype=text&itemID=NRO-DD.SK.218.1&pageseq=1.

Robert FitzRoy c. 1836 from Barlow 1967.

for South America and return by way of the Pacific and so circumnavigate the globe. FitzRoy's appeal for a naturalist was passed to his superior, the Hydrographer of the Navy, Francis Beaufort (1774–1857). Beaufort then appealed to his contact at Cambridge, George Peacock. Peacock thought first of Henslow who was very tempted but was unable to accept. Henslow next suggested Jenyns who was also tempted but equally unable to accept because of his own duties. Henslow and Jenyns were in no doubt as to who to recommend for the post — the recently graduated and scientifically accomplished Darwin. For many years it was believed that Darwin was invited as a gentleman companion to FitzRoy, and was not the official naturalist on the *Beagle*. This has recently been shown to be incorrect. Darwin really was the naturalist on the *Beagle*.[126]

Darwin was delighted with the opportunity and after convincing his initially sceptical father, who would have to foot most of the bill, Darwin set out

[126] See however Wyhe, John van. 2013. "my appointment received the sanction of the Admiralty": Why Charles Darwin really was the naturalist on HMS Beagle. *Studies in History and Philosophy of Biological and Biomedical Sciences* 44 (3): 316–326.

for Cambridge to visit Henslow to discuss the expedition. After hurried preparations in London, Darwin departed on the *Beagle* in December 1831. The *Beagle* voyage is recounted in many works, foremost of which is Darwin's own *Journal of researches* (1839), usually now called *The voyage of the Beagle.*[127] Throughout the voyage Darwin shipped home thousands of geological, botanical and zoological specimens to Henslow in Cambridge. Henslow presented some of Darwin's specimens and read extracts of his letters at meetings of the Cambridge Philosophical Society and these were reported in the press of the day.[128] Henslow even had a collection of Darwin's letters printed, parts of which were also reprinted in the periodical literature.[129] Sedgwick communicated the letters to the Geological Society of London on 18 November 1835 which were also published. This example of the public reading of Darwin's letters and their publication without his knowledge are just one of the countless examples of the common practice of treating unpublished writings on scientific subjects at the time.[130]

---

[127] See Keynes, Richard. 2002. *Fossils, finches and Fuegians : Charles Darwin's adventures and discoveries on the Beagle, 1832–1836.* London: HarperCollin; and Chancellor, Gordon and Wyhe, John van, with the assistance of Kees Rookmaaker (eds). 2009. *Charles Darwin's notebooks from the voyage of the Beagle.* Cambridge: University Press.

[128] Anon. 1835. [Report of a meeting of the Cambridge Philosophical Society]. *The Times* 22 December: 7; Anon. 1836–1837. [Reports of Darwin's communications read to the Cambridge Philosophical Society 1835–1837]. *The London and Edinburgh philosophical magazine and journal of science* 8, no. 43 January 1836: 79, 80; 10, no. 61 April 1837: 316.

[129] Darwin, C. R. [1835]. [Extracts from letters addressed to Professor Henslow]. Cambridge: [privately printed].

[130] Darwin, C. R. 1836. Geological notes made during a survey of the east and west coasts of S. America, in the years 1832, 1833, 1834 and 1835, with an account of a transverse section of the Cordilleras of the Andes between Valparaiso and Mendoza. [Read 18 November 1835] *Proceedings of the Geological Society* 2: 210–212. Further extracts were reprinted in a review in the *Magazine of natural history* 9(64): 441–445 (1836) making it possibly the first review of Darwin's writings. Recently admirers of Alfred Russel Wallace have claimed that it was improper, unethical and even illegal to present and publish Wallace's famous Ternate essay in 1858 without his explicit prior consent. When Darwin found out, he reacted similarly to Wallace, pleased his writing was considered worthy of publication and discussion by his seniors and somewhat embarrassed he had not given the work a final check. Part of the error of the Wallace enthusiasts is the mis-reading of a later letter of Wallace in which he remarked "It was printed without my knowledge, and of course without any correction of proofs. I should, of course, like this act to be stated." The words "printed without my knowledge", to modern readers, mean "without my consent". But when Wallace was writing the phrase meant that others thought so highly of something that it was published without even needing to attempt to publish it onesself. It was an expression of modesty, not of resentment.

HMS *Beagle* in the Galapagos, 17 October 1835 2.15 p.m., oil painting by John Chancellor (1925–1984), 1981. Courtesy of Gordon Chancellor.

After circumnavigating the globe, the *Beagle* returned to England in October 1836. After disembarking Darwin went directly to Shrewsbury to see his family, before heading back to Cambridge to arrange his veritable mountain of specimens in December 1836. At first Darwin considered living in Christ's College and inquired about rooms from the Tutor, Edward John Ash. But as Darwin felt he would probably need to move to London after a few months, the year-long College accommodation arrangement, as well as the need to furnish it and buy crockery etc., was inconvenient. Instead, Adam Sedgwick found lodgings for Darwin at 22 Fitzwilliam Street. The three-storey terraced brick house is marked today with a stone plaque.

Darwin arrived back in Cambridge on 13 December 1836 and stayed with Henslow. On the afternoon of his arrival the diarist and Fellow of Trinity Joseph Romilly (1791–1864) visited the Henslows, as he recorded in his diary: "Drank tea with Mrs Henslow to meet Marchesa &c: here also I met Mr Darwin (g[rand]. s[on]. of [the author of the] Bot^c^ Garden [1789–1791]) who is just returned from his travels round the world: he declares that in 'terra del fuego' whenever a scarcity occurs (wch is every 5 or 6 years) they kill the old women as the most useless living creatures: in conseq. when a famine begins the old women run away into the woods & many of them perish miserably there..."[131] Darwin told some of this story in the second edition of his *Journal of researches*: "when pressed in winter by hunger, they kill and devour their old women before they kill their dogs: the boy, being asked by Mr. Low why they did this, answered, 'Doggies catch otters, old women no.'"[132]

Darwin told similar voyager's tales at Christ's, where physician George Edward Paget (1809–1892) was also dining. Paget later recalled:

> I met [Darwin] at dinner in the rooms of Ash, who was then Tutor in the College. ... He told me also that he had tested in the market at Lima the acute faculty of smell at that time commonly attributed to Vultures. There was a row of these birds for sale I think. [Darwin] walked before them and past them with a large piece of meat in his pocket. The birds took no notice whatever. He then threw within their reach the piece of meat wrapped up in paper. Still the birds

[131] Bury, J. P. T. 1967. *Romilly's Cambridge diary 1832–42.* Cambridge: University Press, p. 110.

[132] Darwin, C. R. 1845. *Journal of researches into the natural history and geology of the countries visited during the voyage of H.M.S. Beagle round the world, under the Command of Capt. Fitz Roy, R.N.* 2nd edition. London: John Murray, p. 214.

Darwin's lodgings at 22 Fitzwilliam Street. The stone plaque, presumably erected in 1909, reads "Charles Darwin lived here 1836–7".

> took no notice. But when with the end of a stick, he uncovered a part of the meat they instantly rushed to seize it.[133]

Being back at Christ's after all of his adventures was a strange feeling. Darwin wrote to Fox on 15 December: "It appears to me, most strange to stand in the

[133] Paget, G. E. 1882. [Recollections of Darwin, 13 September.] (DAR 112.A86-A91), transcribed by Kees Rookmaaker. *The Complete Work of Charles Darwin Online* (http://darwin-online.org.uk/). Darwin's wrote of this experiment in *Birds*, pp. 5–6.

court of Christ, and not to know one undergraduate: It was however some kind of satisfaction to find all the old 'gyps'." On 16 December, Darwin moved into lodgings in Fitzwilliam Street. He continued to dine at Christ's. The College wine book demonstrates that he dined at Christ's on 19 October during a short stay in Cambridge and was possibly listed as "MA" for 26 December and perhaps two other entries for an MA around this time. In one instance the nib was even fined for not naming MA. The fine was later crossed out. On the 29th, "Mr Darwin [was fined for being] too late in hall." He paid his fine with a half bottle of port on each of the two following evenings. On 23 February 1837 Darwin bet one of the Fellows, Edward Baines (1801–1882), that he could guess the height of the ceiling in the Combination Room (now the Old Combination Room) where the Fellows retired for drinks and conversation after dining. The room was overlooked by a portrait of the Foundress of the College, Lady Margaret, and a full length portrait of William Paley.

> 23 Feb. 1837. Mr Darwin *v.* Mr Baines. That the Combination Room measures from the ceiling to the floor more than *(x)* feet.
>
> N.B. Mr Darwin may measure at any part of the room he pleases.[134]

Darwin's name is crossed through, which means he lost the bet. It was forbidden to bet for money, so bets were always for a bottle of port. Indeed if the person laying a bet forgot to say it was for a bottle, he could be fined a bottle! Interestingly, Baines bet the height of the Combination Room ceiling "is nearer (x) than (y) feet" on 5 February 1833, so he must have known he would beat the world traveller.

Darwin recorded in his pocket diary or Journal "Jan [1837]: Cambridge — time spent in arranging general collection; examining minerals, reading, & writing little journal [of Researches] in the evenings Paid two short visits to

[134] F. Darwin 1887. *The life and letters of Charles Darwin.* London: Murray, vol. 1: 279. The Wine Book remains one of the most interesting traces of Darwin's presence in the College and is regularly taken from its place in the row of quarter leather bound volumes in the cabinet in the Combination Room to show to interested visitors. The Wine Book's contents are usually considered to be secret and not to be shared outside the Room. I have chosen, therefore, to cite the 1887 published transcription of Francis Darwin rather than reproduce a photograph of the original book. See the interesting account of the Wine Book from an earlier period, by Steel, Anthony. 1949. *The custom of the room or early wine-books of Christ's College.* Cambridge.

London. — & read paper on elevation of coast of Chile." On 27 February, he presented a paper at a meeting of the Cambridge Philosophical Society. The minutes of the General Meeting record: "An account by Mr C. Darwin of fused sand tubes found near the Rio Plata, which were exhibited along with several other specimens of rocks." Darwin wrote to his sister Caroline: "I have just been reading a short paper to the Philosoph. Soc^y^. of this place, and exhibiting some specimens & giving a verbal account of them. It went off very prosperously & we had a good discussion in which Whewell & Sedgwick took an active part. … On Friday morning I migrate [to London]. My Cambridge life is ending most pleasantly."[135] Apparently on 6 March 1837, Darwin moved to London to be closer to the scientific colleagues and institutions who were discussing his *Beagle* collections.[136] Darwin wrote to Fox from London "The only evil I found in Cambridge, was its being too pleasant".[137] Darwin's formative life of living in Cambridge had come to an end.

[135] *Correspondence*, vol. 2: 8–10.

[136] *Journal*.

[137] *Correspondence*, vol. 2: 11.

THE ORIGIN OF SPECIES AND HONORARY CAMBRIDGE DEGREE

The plan to become a clergyman, as Darwin later recollected, was never "formally given up, but died a natural death when on leaving Cambridge I joined the *Beagle* as Naturalist."[138] But for a brief time at least he imagined spending the rest of his life living and working in Cambridge. In a memorandum to himself written around April 1838, Darwin weighed up the pros and cons of marriage. If he married, he feared, he would have to earn a living to adequately support a wife and family. The first possibility that came to mind was a: "Cambridge Professorship, either Geolog, or Zoolog." As he teetered towards the inevitability of marriage he decided there were two alternatives. "Cambridge, better, but [it would be like being a] fish out of water, not being Professor & poverty. Then Cambridge Professorship, — & make best of it, do duty as such & work at spare times — My destiny will be Camb. Prof. or [to be a] poor man; [and live on the] outskirts of London, some small Square &c: — & work as well as I can".[139]

Darwin married his cousin Emma Wedgwood in 1839 and the money settled on him by his father and new father-in-law was more generous than Darwin had

---

[138] *Autiobiography*, p. 57.

[139] Darwin, C. R. 1838. 'Work finished If not marry' [Memorandum on marriage]. (DAR210.8.1), transcribed by Kees Rookmaaker. *The Complete Work of Charles Darwin Online. Darwin Online*, http://darwin-online.org.uk/content/frameset?keywords=finished%20work&pageseq=2&itemID=CUL-DAR210.8.1&viewtype=side.

expected. At first they continued to live in London and from 1842 in a quiet country house at Down, Kent — still known as Down House. Darwin spent the next nineteen years publishing the results and collections of the *Beagle* voyage. Early in this process, he had formulated his theory of evolution by natural selection. He worked on it in the background while he pursued his *Beagle* publishing programme and the marine invertebrate work which was its extension. As Darwin wrote in 1857 to the American botanist Asa Gray (1810–1888): "it occurred to me that whilst otherwise employed on Nat. Hist, I might perhaps do good if I noted any sort of facts bearing on the question of the origin of species; & this I have since been doing."[140] This is very far indeed from the frequently repeated traditional view that Darwin was afraid to publish his theory and thus withheld it or that he kept it secret. I have elsewhere argued that, just as Darwin describes in this letter, he proceeded to work full-time on his unfinished theoretical species programme once he had finished earlier projects. There was no postponement or conscious delay.[141]

After twenty years of tireless scientific publishing and research, Darwin published his life's work, *On the origin of species by means of natural selection, or the preservation of favoured races in the struggle for life*, in November 1859. Its importance in altering our understanding of life on Earth is difficult to exaggerate. Between the covers of a single volume, Darwin managed to demonstrate the most fundamental patterns of life on Earth. At a stroke all of the families, genera and countless thousands of species were all connected in one single and beautifully simple system. All life is related, genealogically, on a great branching tree, the tree of life. He called it 'the theory of descent with modification through natural selection'.

Darwin began by explaining how he came to doubt the stability of species and how long he had worked on the subject. The brute facts of the similarities of different species, the similarities during embryological development of members of the same genus, geographical distribution, the progressive succession of fossil forms and so forth showed that species change. But Darwin also showed how they changed and, most importantly, how they came to be so wonderfully adapted to their environments, and their immensely complex relationships with

---

[140] Darwin to Gray 20 July [1857], *Correspondence*, vol. 6: 432.

[141] Wyhe, John van. 2007. Mind the gap: Did Darwin avoid publishing his theory for many years? *Notes and Records of the Royal Society* 61: 177–205 and Wyhe, J. van. 2013. *Dispelling the Darkness: Voyage in the Malay Archipelago and the discovery of evolution by Wallace and Darwin.* Singapore: World Scientific Press.

Pencil drawing of Darwin, 1839 by George Richmond. Cambridge University Library.

one another. Natural selection explains how adaptation occurs, over many generations, given the commonly accepted, but often overlooked, properties of living things. The conclusion to the *Origin of Species* is still stirring today:

> ...all living things have much in common, in their chemical composition, their germinal vesicles, their cellular structure, and their laws of growth and reproduction. ...There is grandeur in this view of life, with its several powers, having been originally breathed into a few forms or into one; and that, whilst this planet has gone cycling on according to the fixed law of gravity, from so simple a beginning endless forms most beautiful and most wonderful have been, and are being, evolved.[142]

[142] Darwin, C. R. 1859. *On the origin of species by means of natural selection, or the preservation of favoured races in the struggle for life*. London: John Murray, p. 490.

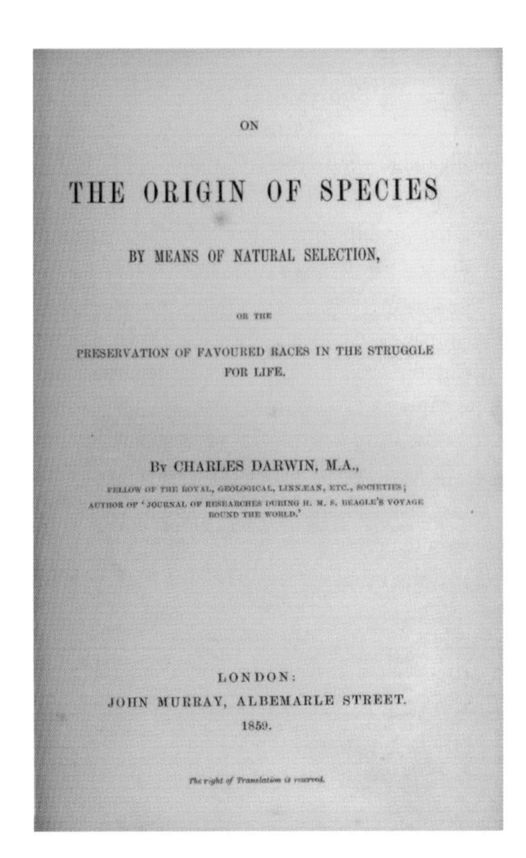

ON

THE ORIGIN OF SPECIES

BY MEANS OF NATURAL SELECTION,

OR THE

PRESERVATION OF FAVOURED RACES IN THE STRUGGLE
FOR LIFE.

By CHARLES DARWIN, M.A.,

FELLOW OF THE ROYAL, GEOLOGICAL, LINNÆAN, ETC., SOCIETIES;
AUTHOR OF 'JOURNAL OF RESEARCHES DURING H. M. S. BEAGLE'S VOYAGE
ROUND THE WORLD.'

LONDON:
JOHN MURRAY, ALBEMARLE STREET.
1859.

The right of Translation is reserved.

Title page of the first edition of the *Origin of Species* (1859).

The book was immediately controversial and widely discussed. The reactions to Darwin's evolutionary theories were varied and pronounced. To many it seemed irreligious; to others unscientific. In zoology, taxonomy, botany, palaeontology, philosophy, anthropology, psychology, literature and religion, Darwin's work engendered profound reactions — many of which are still ongoing. However, the remarkable fact is in less than twenty years the international scientific community had accepted evolution is a fact. It has never again been a matter of scientific debate.

Darwin's friend and the co-discoverer of natural selection, Alfred Russel Wallace, explained at the outset of his great book *Darwinism* (1889):

> Before Darwin's work appeared, the great majority of naturalists, and, almost without exception the whole literary and scientific world, held firmly to the belief

> that *species* were realities, and had not been derived from other species by any process accessible to us; the different species of crow and of violet were believed to have been always as distinct and separate as they are now, and to have originated by some totally unknown process so far removed from ordinary reproduction that it was usually spoken of as "special creation." There was, then, no question of the origin of families, orders, and classes, because the very first step of all, the "origin of species," was believed to be an insoluble problem. But now this is all changed. The whole scientific and literary world, even the whole educated public, accepts, as a matter of common knowledge, the origin of species from other allied species by the ordinary process of natural birth. The idea of special creation or any altogether exceptional mode of production is absolutely extinct! Yet more: this is held also to apply to many higher groups as well as to the species of a genus, and not even Mr. Darwin's severest critics venture to suggest that the primeval bird, reptile, or fish must have been "specially created." And this vast, this totally unprecedented change in public opinion has been the result of the work of one man, and was brought about in the short space of twenty years![143]

By the 1870s Darwin had consequently become one of the most famous and revered living naturalists. On 17 November 1877, he was awarded an honorary LL.D. (Doctor of Laws) degree by the University of Cambridge. The economist Neville Keynes (1852–1949) witnessed the ceremony and recorded in his diary:

> The honorary degree of LLD was conferred upon Darwin in the Senate House amidst a scene of some disorder. The building was crammed, floor and galleries, the undergraduates being chiefly in the galleries; and it was of course an occasion on wh. undergraduate wit felt bound to distinguish itself. The chief pleasantry consisted of a monkey swung across by strings from gallery to gallery, which monkey was in the course of the proceedings to have been changed into a man. Before however this desirable consummation was reached, the representative of the original ancestor, (than whom he was less fortunate), was seized by one of the proctors, & thus prevented from fulfilling his high destiny. The perpetrators of the joke were very wrath, & vented their fury chiefly in groans for Humphreys, the most unpopular of the proctors. He was also made the butt of such remarks as this, Would Dr. Darwin kindly afford us some information regarding the ancestors of Mr. Humphreys? — a sally which took amazingly. Sandys, the public orator, introduced Darwin, according to custom, in a rather

[143] Wallace, A. R. 1889. *Darwinism: an exposition of the theory of natural selection with some of its applications*. London & New York: Macmillan & Co, pp. 8–9.

long Latin oration, wh. was delivered amidst a ceaseless fire of interruptions, (chiefly feeble), from the wittiest of the undergraduates. Sandys (I imagine inadvertently) made use of the word apes, & then the cheering was enormous. Darwin bore himself in a rather trying position with remarkable dignity; but I heard afterwards that his hand shook so much while he was signing the registry, that his signature was scarcely legible. Another emblem swung from the galleries was a large ring of iron, adorned with ribbons, supposed to represent the missing link. It was ultimately swung down into the lap of one of the lady visitors, who pluckily cut it down and appropriated it, amidst tremendous applause.[144]

Interior of Senate House.

[144] Keynes, N. 1877. [Recollection of Darwin's honorary LLD degree]. Diary. (CUL-Add.7831.2), transcribed by Kees Rookmaaker. *The Complete Work of Charles Darwin Online* (http://darwin-online.org.uk/). The Latin oration (CUL-DAR139.2.1) was printed and is available on *Darwin Online*: http://darwin-online.org.uk/content/frameset?itemID=CUL-DAR139.2.1&viewtype=image&pageseq=1. An English translation is also on *Darwin Online*: http://darwin-online.org.uk/content/frameset?pageseq=1&itemID=CUL-DAR140.1.13&viewtype=image.

A photograph of the stuffed monkey that was suspended above Darwin at Senate House. The monkey, wearing a cap and gown, is sitting on copies of Darwin's *Descent of Man* (1871), *Expression of the Emotions* (1872) and *Origin of Species* (1859). Courtesy of the Old Library of Christ's College, Cambridge. This photograph was part (item 248) of the Darwin exhibition at Christ's in 1909 and was "Lent by Dr John E. Marr, F.R.S.".

Darwin's wife, Emma (1808–1896), described the same event in a letter to her son William from a Darwin family perspective:

> There seemed to be periodical cheering in answer to jokes which sounded deafening; but when [Darwin] came in, in his red cloak, ushered in by some authorities, it was perfectly deafening for some minutes.... We had been watching some cords stretched across from one gallery to another wondering what was to happen, but were not surprised to see a monkey dangling down which caused shouts and jokes about our ancestors, etc. ... At last the Vice-Chancellor appeared, more bowing and hand-shaking, and then [Darwin] was marched down the aisle behind two men with silver maces, and the unfortunate Public

Orator came and stood by him and got thro' his very tedious harangue as he could, constantly interrupted by the most unmannerly shouts and jeers... At last he got to the end with admirable nerve and temper, and then they all marched back to the Vice-Chancellor in scarlet and white fur, and [Darwin] joined his hands and did not kneel but the Vice-Chancellor put his hands outside and said a few Latin words, and then it was over, and everybody came up and shook hands....I felt very grand walking about with my LL.D. in his silk gown.[145]

Darwin later wrote to Fox on 2 December 1877: "We had a grand time of it in Cambridge & I saw my old rooms in Christ's where we spent so many happy days."[146]

In 1878, Darwin made a rare contribution to Cambridge University reform. He co-signed a memorial to the Vice-Chancellor proposing that candidates for honours should no longer be required to pass an examination in Greek.[147] The memorial, also reprinted in *The Times*, was only one of many in a long and circuitous debate about University reforms. In 1881 Darwin made a donation, heretofore apparently overlooked by historians, towards the construction of a physical and biological laboratory for women in Cambridge.[148]

Inspired by the award of Darwin's honorary degree, meetings were held in the Combination Room at Christ's College to propose a memorial to Darwin.[149] After £400 was raised by subscription, Darwin's portrait was painted in oils by William Blake Richmond (1842–1921) in June 1879. It now hangs outside the Zoology Department Library. Emma Darwin called it "the red picture" (because of the scarlet LL.D. gown), "I thought it quite horrid, so fierce and so dirty".[150]

Darwin died at his home, Down House, Kent, on 26 April 1882. There was no deathbed conversion or retraction of his theories as sometimes alleged by conspiracy theorists.[151] Darwin was buried after a state funeral in Westminster

---

[145] Litchfield, H. E. (ed.) 1915. *Emma Darwin: a century of family letters, 1792–1896*. London: John Murray, vol. 2: 230–231.

[146] Christ's College Library, Fox 155.

[147] Darwin, C. R. *et al.* 1878. [Memorial to the Vice-Chancellor respecting the Examination in Greek in the Previous Examination]. *Cambridge University Reporter* (7 December): 206–207.

[148] Anon. 1881. The scientific education of woman [with a donation by Darwin for a Cambridge laboratory for women]. *The Times* 27 January: 4.

[149] See Anon. 1879. University intelligence [meeting in Christ's College, Cambridge of the committee of the Darwin Memorial Fund]. *The Times* 8 March: 9.

[150] Litchfield, H. E. (ed.) 1915. *Emma Darwin: a century of family letters, 1792–1896*. London: John Murray, vol. 2: 248.

[151] See Moore, J. R. 1994. *The Darwin legend*. Grand Rapids, Michigan: Baker Books.

Abbey. An unknown writer at Christ's College amended the College record book 'Admissions 1815–1852' (T.3.1): "Name off April 1882 Dead", thus, somewhat coldly, finally signing Darwin off the books as a member of Christ's College.

The "red picture" by William Blake Richmond. Department of Zoology, University of Cambridge.

The international reaction to Darwin's death was unprecedented. Hundreds of obituaries were published around the world. Reading these today, it is striking how consistent they are. The reason this is surprising is that so many writers from so many different backgrounds wrote so highly of Darwin as a person, as a man of science and as the figure who far more than all others, had changed the entire face of the life sciences forever. This was before biographies and history books had the chance to disseminate particular versions of Darwin's life. What emerged in the obituaries was the almost simultaneous effusion of a generation — a generation which had read and experienced first-hand the impact of Darwin's astonishingly innovative scientific work. Many writers declared that Darwin had effected a revolution in our understanding of nature unrivalled by any thinker since Newton, or as others put it, unrivalled in any age. The popular science writer Grant Allen wrote in *The Academy*:

> In 1859, the *Origin of Species* at last appeared, under what circumstances all the world knows. It was nothing less than a revolution; it marks the year 1 of a new era, not for science alone, but for every department of human thought — nay, even of human action. ... the influence of his thought upon the thought of the age has far outweighed any influence ever before exerted by a single man during his own lifetime. He has revolutionised, not biology alone, but all science; not science alone, but all philosophy; not philosophy alone, but human life. Man, his origin and nature, his future hopes and realisable ideals, all seem something different to the present generation from their seeming to the generations that lie behind us in the field of time.

Later in 1882, revised statutes for Christ's College passed through Parliament. One of the new changes allowed the College to appoint Honorary Fellows, and it had been the intention of the Governing Body to confer this title on Darwin. In 1883, the artist William Walter Ouless (1848–1933) painted a copy of his 1875 portrait of Darwin which had been commissioned by the Darwin family as a birthday present for Darwin. The copy now hangs in the Hall of Christ's College, facing a portrait of William Paley. The original Ouless portrait hangs in Darwin College, Cambridge. Francis Darwin wrote: "Mr. Ouless's portrait is, in my opinion, the finest representation of my father that has been produced".[152] Although Darwin wrote to a friend that he thought "I look a very venerable, acute, melancholy old dog; whether I really look so I do not know".[153]

In 1882 and in following years stained glass portraits by Messers. Burlison and Grylls were added to the Hall's west oriel.[154] Darwin is depicted in the top right and final panel in his scarlet honorary LL.D. gown facing, once again, William Paley.

In 1885 Darwin's son George (1845–1912) donated a Wedgwood portrait medallion in green jasper to be set into the north wall in the sitting room of Darwin's old College rooms. A second jasper panel beneath Darwin's portrait reads "CHARLES ROBERT DARWIN 1829–31". The date 1829 is a mistake for 1828. A small label on the underside of the frame reads "Erected by G. H. Darwin. Plumian Professor. 1885."

---

[152] F. Darwin 1887. *The life and letters of Charles Darwin.* London: Murray, vol. 3: 95.

[153] F. Darwin 1887. *The life and letters of Charles Darwin.* London: Murray, vol. 3: 195.

[154] Peile, J. 1900. *Christ's College.* London: Robinson, p. 16.

The Paley and Darwin panels in the stained glass window of the Hall of Christ's College.

"A very venerable, acute, melancholy old dog". Oil painting of Darwin by William Walter Ouless (1883) which hangs in the College Hall.

The framed Wedgwood medallion (1885) in the wall of Darwin's rooms.

The Darwins at Christ's College

Darwin sent his eldest son William Erasmus (1839–1914) to Christ's in 1858 where he was admitted pensioner under Messrs Hays and Gunson on 13 March. After his first Michaelmas Term, William lived in his father's old rooms and even decorated them with some of the prints his father had proudly displayed there in the 1820s and 1830s. William found that Impey, Darwin's old gyp, was still there. Darwin wrote to Fox: "William, my son, is now at Christ Coll. in the rooms above yours. My old Gyp. Impey was astounded to hear that he was my son & very simply asked "why has he been long married?" What pleasant hours, those were when I used to come & drink coffee with you daily!"[155] William earned his B.A. in June 1862 and later became a banker in Southampton.

---

[155] *Correspondence*, vol. 7: 196. See Darwin, F. 1914. [Obituary of] William Erasmus Darwin. *Christ's College Magazine* 29: 16–23.

Postcard of Christ's College c. 1860s.

Darwin's son Francis (1848–1925) studied at Trinity College and became a Fellow of Christ's on 8 December 1888, and Honorary Fellow in 1906. Darwin's grandson, Sir Charles Darwin (1887–1962), was the 29th Master of Christ's College from 1936–1939. Two of Darwin's other sons made their lives in Cambridge. George Darwin (1845–1912) was educated at Trinity College (B.A. 1868) and became Plumian Professor of Astronomy and Experimental Philosophy at Cambridge in 1883. He purchased Newnham Grange in 1885. His daughter, Gwen Raverat (1885–1957), wrote a delightful account of growing up at Newnham Grange in *Period piece: A Cambridge childhood* (1952). Darwin College was founded on the site in 1964. Horace Darwin (1851–1928) was also educated at Trinity College, Cambridge. He received his B.A. in 1874. He became the Founder and Director of the Cambridge Instrument Co., Botolph Lane, Cambridge and was Mayor of Cambridge between 1896 and 1897. Charles Galton Darwin (1887–1962), a grandson, was Master of Christ's between 1936 and 1939. Descendants of Darwin remain members of the University of Cambridge to this day.

Because of Charles Darwin's attachment to Cambridge and that of his sons, it is hardly surprising that much of his library and papers found their way to Cambridge for long-term preservation and study by scholars. After the death of Darwin's widow, Emma, in 1896, a portion of their library was donated to the

University Library in 1900.[156] The majority of Darwin's scientific library was donated to the Botany School, Cambridge in 1908.[157] (The collection is now kept in the Cambridge University Library Rare Books Room.) Most importantly of all, Darwin's private papers were donated to the University by the Pilgrim Trust and the Darwin family in 1942.[158]

---

[156] Anon. 1900. List of donations [books] received during the year 1899: From the executors of the late Mrs Darwin. *Cambridge University Reporter* 30 41. 15 June: 1079–1080.

[157] Rutherford, H. W. 1908. *Catalogue of the library of Charles Darwin now in the Botany School, Cambridge*. Compiled by H. W. Rutherford, of the University Library; with an Introduction by Francis Darwin. Cambridge: Cambridge University Press.

[158] Barrett *et al.* 1987, p. 2. 1960. *Handlist of Darwin papers at the University Library Cambridge*. Cambridge: University Library. See the Darwin Manuscript Catalogue on *Darwin Online*, http://darwin-online.org.uk/MScatintro.html.

1909: THE FIRST DARWIN CENTENARY IN CAMBRIDGE

On 22–24 June 1909, over 400 scientists and dignitaries from 167 different countries gathered at Cambridge to celebrate the centenary of Darwin's birth and the fiftieth anniversary of the publication of the *Origin of Species*. The event was an unprecedented success — never before had such a celebration been held, not for an institution or a nation — but for an individual scientist.[159] The celebrations began with a reception in the Fitzwilliam Museum. On 23 June presentations and speeches in the Senate House were followed by a garden party given by the Master and Fellows of Christ's College in the College grounds.

In the evening a banquet was held in the New Examination Hall of the University, the Museums Site. During the banquet Darwin's eldest son William (1839–1914) gave one of the speeches. He remarked: "I am sure my father would have said, though, perhaps, with a tone of apology in his voice, that if there was to be a [Darwin] celebration there could be no more fitting place than Cambridge. He always retained a love for Cambridge and a happy memory of his

[159] See Richmond, Marsha. 2006. The 1909 Darwin celebration: reexamining evolution in the light of Mendel, mutation, and meiosis. *Isis*. 973: 447–484. A useful collection of publications from the 1909 events are reproduced in *Darwin Online*: http://darwin-online.org.uk/1909.html. See also Smocovitis, V. B. 1999. The 1959 Darwin Centennial Celebration in America. *Osiris* 14: 274–323.

Photographs of the garden party held at Christ's College on 23 June 1909. From the Shipley album in the Old Library of Christ's College.

life here. It was the happiest and gayest period of his life".[160] The banquet was followed with a reception by the Master and Fellows of nearby Pembroke College. On the 24th honorary degrees were conferred in the Senate House, and Sir Archibald Geikie, President of the Royal Society, gave his Reed lecture 'Darwin as geologist'.[161] In the afternoon, there was a garden party at Trinity College given by Darwin's children.

Darwin's rooms at Christ's were open to visitors during the afternoon of the 23rd and the morning and afternoon of the 24th. They were probably tidied up prior to this opening. On this occasion, the sitting room was photographed by Cambridge photographer John Palmer Clarke. For one hundred years, this was almost the only photograph of Darwin's rooms in print. It was often assumed to show what Darwin's rooms looked like in his own day, when in fact the furnishings are simply those of the current occupant, a Fellow, in 1909.

Darwin's sitting room photographed in 1909.

An historic exhibition of Darwiniana was held in the downstairs room of the Old Library at Christ's. It included portraits, busts, notebooks, manuscripts, letters and artefacts used on the *Beagle* voyage. Most were lent by

[160] Darwin, G. and Darwin, F. (eds.) 1909. *Darwin celebration, Cambridge, June, 1909. Speeches delivered at the banquet held on June 23rd.* Cambridge: Cambridge Daily News, p. 15.

[161] Geikie, A. 1909. *Charles Darwin as geologist: The Rede Lecture given at the Darwin Centennial Commemoration on 24 June 1909.* Cambridge: University Press.

Photograph of the Darwin exhibition at Christ's College in 1909. From the Shipley album in the Old Library.

Darwin's children. Some of the *Beagle* notebooks and Darwin's transmutation notebooks were first displayed here. The printed exhibition catalogue listed 257 items.[162] Many distinguished visitors from around the world came to see the exhibition and signed their names in a book now kept in the Old Library.

Some of the objects on display remained with the College or were later returned to it. The bronze medallion visible in the photograph above the chair was by the sculptor Horace Montford (1840–1919) who made the sculpture of Darwin that stands in front of Shrewsbury school. The medallion was listed in the exhibition catalogue as item number 182 and that it was lent by Montford. In 2009, the medallion was found in a College storeroom and mounted inside the great gate of the College where it is visible to the public.

---

[162] Shipley, A. E. and Simpson, J.C. (eds.) 1909. *Darwin centenary: the portraits, prints and writings of Charles Robert Darwin, exhibited at Christ's College, Cambridge 1909.* [Cambridge: University Press].

Photographs of the Darwin exhibition at Christ's College in 1909. From the Shipley album in the Old Library.

Another item still in the College is a smartly framed photograph of Darwin hung on the end of a book case in the Old Library for over seventy years. The back of the frame reveals that it was made by a local cabinet maker on Trumpington Street and that this was item 184 in the exhibition "7th left". It is visible in the middle photograph above on the second bookcase. The photograph, and the frame, have their own story. A pencilled note on the wooden back of the frame explains: "This photograph of Darwin was presented by him to my Uncle, F.D. Dyster, of Tenby. I am informed by Francis Darwin, his son, that the photograph was probably taken in the year 1854, but he had never seen it. F. H. H. Guillemard." Below this, in a different and very faint hand is written in pencil: "NB F.D. Dyster was the microscopist after whom the genus Dysteria was named." A third pencil note in yet a different hand records: "Exhibited at the Darwin Commemoration in Christ's College — June 1909." Finally, a typewritten slip pasted on the board states: "CHARLES DARWIN (probably 1854, aged 45). Given by him to Dr. F.D. Dyster, and bequeathed to the College by Dr. Dyster's nephew, Dr. F.H.H. Guillemard, 1934." There is an entry for the photograph in the 1909 exhibition catalogue.

> 91. PORTRAIT OF CHARLES DARWIN.
> Lent by Dr. F. H. H. Guillemard.
> Photograph probably taken about 1854 and given by Charles Darwin to F. D. Dyster, Esq., the microscopist.[163]

The photograph was taken in London c. 1854, about one year after Darwin started full-time work on his species theory, by Maull and Polyblank for the Literary and Scientific Portrait Club. A letter from Darwin to Joseph Dalton Hooker on 27 May 1855 refers to this photograph: 'if I really have as bad an expression, as my photograph gives me, how I can have one single friend is surprising.'[164]

There is no known surviving Darwin correspondence with the man Darwin gave the photograph to, Frederick Daniel Dyster (1810–1893). He was a surgeon naturalist with strong interests in microscopy and marine zoology. Dyster was a

---

[163] A. E. Shipley, *Darwin Centenary. The Portraits, Prints and Writings of Charles Robert Darwin. Exhibited at Christ's College, Cambridge, 1909.* [Privately printed].

[164] *Correspondence*, vol. 5: 339.

friend of Thomas Henry Huxley and attended Emma Darwin's aunt, Jessie Sismondi, when she died on 3 March 1853.[165]

Also inside the frame, just below the photograph, a small rectangular opening in the mount displayed the signature 'Ch. Darwin'. I was curious to know if this signature was just a scrap of paper cut from a letter, or if it was written on a document still preserved, though sealed inside the 35.5×40 cm frame. The College Librarian, Candace Guite, and the Keeper of Pictures, David Norman, gave permission for the frame to be removed and opened. This was done by Conservation Officer, Melvin Jefferson.

On opening the frame, which had presumably been sealed since 1909, Jefferson found the Darwin signature to be the endorsement on the back of a cheque. The entire cheque had been carefully folded and preserved so that just the signature on the back could be seen through the mount.

The cheque is from the Union Bank of London, made out by Darwin 'to self' for 100 pounds on 21 March 1872.

The entire cheque transcribed is as follows (Darwin's writing is in bold):

No. V18356 London **March 21** 18**72**
The Union Bank of London,
CHARING CROSS BRANCH 4, PALL MALL EAST.
Pay **to self** or Order **One Hundred Pounds**
100.0.0
This Cheque must be endorsed by the party to whom it is payable **Ch. Darwin**

On the reverse it is signed: 'Ch. Darwin' — this being the signature visible under the photograph. The cheque is stamped: 'Paid Mar. 22' A hole in the centre probably indicates Darwin put the cheque on his spits.

What was Darwin doing on 21 March 1872? We know from his 'Journal' or diary that he and his family left their rented London holiday house on that very day to return to their home in the village of Downe, Kent.[166]

---

[165] Litchfield, H.E. (ed.) 1915. *Emma Darwin, A Century of Family Letters, 1702-1896*. London: John Murray, vol. 2, p. 152.

[166] *Journal*: Wyhe, J. van (ed.) 2005. Darwin's personal 'Journal' (1809–1881). (DAR 158), transcribed by the Correspondence of Charles Darwin project. *The Complete Work of Charles Darwin Online* (http://darwin-online.org.uk/).

So how did the cheque end up inside the frame with the photograph Darwin gave to Dyster? Dyster's nephew, Dr Francis Henry Hill Guillemard (1852–1933), a geographer and travel writer, signed the exhibition guest book at Christ's College with Francis Darwin on 11 June 1909. Perhaps Francis Darwin offered the cheque from his father's papers as an example of his signature. Darwin's papers had now become so precious that the cheque was preserved, and the signature not just cut out. The fact that the frame was made in Cambridge is consistent with this.

The framed Darwin photograph and signature in the Old Library.

And so the photograph, and the cheque bearing Darwin's signature, were sealed behind glass in 1909. After the death of Guillemard in 1934, the frame was given to Christ's where it has hung ever since. Both the photograph and the cheque have now been conserved. Excellent reproductions are displayed in the original frame in its original position.

Among the manuscripts on display were "197. MS. notes made by Charles Darwin while reading Paley's *Evidences of Christianity*. Lent by Francis Darwin". I am not aware of the existence or location of these notes which no doubt be of considerable interest. Of less interest is a manuscript draft leaf of Darwin's book *Insectivorous Plants* (1875), p. 427, the verso was later used by George for math ematical notes. This was also lent by Francis Darwin and displayed in a frame as item

The photograph after removal from the frame in 2009.

number 161. This remains in the Old Library to this day.[167] Other items such as the print of the College chapel and the photograph of the monkey are reproduced above.

Also in 1909, an American delegation presented the College with a large bronze bust of Darwin made by the sculptor William Couper (1853–1942) for the New York Academy of Sciences in the same year.[168] It was item number 119 in the exhibition. The first bust was given to the American Museum of Natural

[167] It is published on *Darwin Online*, http://darwin-online.org.uk/content/frameset?viewtype=side&itemID=CC-OldLibraryGG.1.25&pageseq=1.

[168] See: Woram, John. n.d. Portraits in the round: busts of Charles Darwin, http://www.galapagos.to/TEXTS/COUPER.HTM.

No V18356 London 18
The Union Bank of London,
CHARING CROSS BRANCH, 4. PALL MALL EAST.
Pay or Order
PAID MAR. 22
This Cheque must be endorsed by the party to whom it is payable.
UNION BANK OF LONDON.

The cheque found inside the framed photograph of Darwin. Photograph by Melvin Jefferson.

History who commissioned a copy from Couper to send to Cambridge. The bust has stood for many years in an arched brick gallery in the College grounds behind 3rd Court. By 2008 it had come to be known, rather tongue-in-cheek, as 'the shrine'. As the anniversary approached, it was felt inappropriate to use this name in public. I proposed we refer to the structure as the portico, which is the name now used.

Darwin's library, then kept at the Botany School, Downing Street in Cambridge, could also be viewed by visitors in 1909. The Sedgwick Museum displayed rock specimens collected by Darwin during the voyage of the *Beagle*, and the University Library offered an exhibition of manuscripts and books illustrating the progress of science. The Master, Arthur Shipley (1861–1927), compiled a magnificent album containing many of the press cuttings, invitations, menus, speeches and other memorabilia of the 1909 celebrations. This can now be viewed in the Old Library.

An anonymous writer describing the garden party held at Christ's, with its guests from all around the world in their brightly-coloured gowns, wrote a fitting epitaph for the 1909 celebrations which could serve almost as well for the vast amount of activity in the bicentenary year of 2009:

> It was indeed interesting at the time to reflect, while looking upon this richly arrayed throng, that the past energies of one great life had occasioned these general rejoicings, symbolising as they did the honour to the personality of the great naturalist himself and the expectation of greater things to come by the continuance of his labours.

Bronze bust of Darwin by William Couper, 1909. Photographed before the restoration of the plinth and portico in 2009.

2009: THE SECOND DARWIN CENTENARY IN CAMBRIDGE

The year 2009 was the 150th anniversary of the publication of *The Origin of Species* and the 200th anniversary of Darwin's birth. It was by far the largest world-wide commemoration of a historic scientist the world has ever seen. Probably over 1,000 publications, lectures, conferences, books articles documentaries and other commemorative events were created. My own list on *Darwin Online* records over 600.[169] I myself published four books, about eleven articles, gave countless radio, TV, podcast, newspaper and magazine interviews and gave over 40 public lectures in a dozen countries. I expect other Darwin scholars experienced similar demand.[170]

To prepare for the big year of the College's most famous alumnus, a Darwin Committee was formed at Christ's College in 2006. As the Darwin historian of the College, I naturally had a place on this committee and was able to propose some of the actions that were taken. I suggested that we refurbish the Darwin Portico. I wrote text for panels and provided illustrations. Apart from directing the restoration of Darwin's rooms, my major project was researching and writing *Darwin in Cambridge* (2009). The College also opened a wonderful exhibition in

[169] http://darwin-online.org.uk/2009.html.

[170] See Wyhe, John van. 2010 Commemorating Charles Darwin. *Evolutionary review* 1 (1)(February): 42–47.

the Old Library "Charles Darwin: On land and at sea", featuring most of the College's Darwin items and many other interesting objects and memorabilia. The catalogue can be accessed online.[171] This was mostly the work of librarians Candace Guite and Colin Higgins. Members of the College student Darwin Society prepared an educational website: "Charles Darwin & Evolution 1809–2009."[172]

The committee also commissioned a life-sized bronze statue of Darwin as a student by the sculptor and alumnus Anthony Smith. The resulting statue was unveiled by HRH The Duke of Edinburgh on Darwin's 200th birthday, 12 February 2009. I watched the proceedings from Darwin's rooms. That evening there was a formal fundraising dinner in Hall. The statue has since been placed in The Darwin Sculpture Garden. The plants in the garden were chosen to represent Darwin's voyage around the world on the *Beagle*.

Darwin's "Most Snug & Comfortable Rooms"

As the Darwin historian in College and on the committee, I was keen to restore Darwin's rooms to their early nineteenth-century appearance and open them to the public for the first time since 1909. My proposal was accepted and a budget was agreed by the College's Governing Body. I am grateful to the College for the confidence placed in me by assigning me the responsibility and the budget to restore Darwin's rooms. No drawings or descriptions of Darwin's rooms as he knew them survive. The earliest known depiction of the interior of Darwin's rooms is the photograph taken in 1909. By this time, the panelling was painted a dark woodgrain colour. Thus clues had to be sought in contemporary documents ranging from Darwin's letters and private papers, recollections of his Cambridge friends and contemporary illustrations and descriptions of Cambridge student life in the 1820s–1830s and finally the physical evidence of the rooms themselves.

Darwin's rooms are on the first floor of the south side of the first court of Christ's College, now room G4. The stairway leading up to Darwin's rooms was originally twice as wide as it is today and filled the entire passage. The passage was opened in the mid-twentieth century to give access to the new Undergraduate Library. David Bartram, a consultant for the National Trust, conducted an initial

[171] http://www.christs.cam.ac.uk/sites/www.christs.cam.ac.uk/files/migrated-media/exhibition_catalogue_%28finalupdated%29.pdf.

[172] http://darwin200.christs.cam.ac.uk/pages/.

inspection of the rooms with me in 2006 and suggested that the oak panelling around the fireplace, because the wall space is asymmetrical, was built in situ. The remaining panelling was taken from another source, probably from within the College, judging from the similarity of the friezes. Numerous rough cuts and unmatching panels (some inserted upside down) also attest to the fact that the remaining panelling was added later. The two ornamental closet-like doors on the west side open onto brick walls. The brick appears to be late, possibly Victorian. There are thin fluted wooden strips on either side of these doors that are machined, they seem to be present in the 1909 photograph. The sash windows were added at some point in the eighteenth century. Around 1899, the ceiling plastering was removed or replaced. The central exposed beam may then have been sheathed with machined and bevelled planks as it is today. The beams are sheathed identically in the Fellows' Parlour next door.

In 1933, Darwin's rooms were occupied by the physicist and novelist C. P. Snow. He is famous for his proposal that there are "two cultures", the sciences and the humanities, in Western intellectual life. His ashes now rest in an urn by the bathing pool. The rooms then underwent the following renovations.

> The oak panelling which had been covered by many layers of paint, was cleaned, disclosing the woodwork in excellent condition, and at the top, on the carved frieze, underneath all the paint, bright colouring in blue, yellow and red. This colouring, in addition to being an interesting antiquarian relic, is a distinct asset to the appearance of the room. In some places where the work had failed, it has been slightly and carefully restored; the general effect must now be very much the same as it was when the colours were first put up.
>
> Another discovery made at the same time was that of the old clunch fireplace arch behind the modern fittings. The spring of the arch on both sides had been cut away to make room for the later alterations, but the missing parts have now been carefully restored, and the whole effect is extremely handsome.[173]

A few other differences, probably the result of the 1933 renovations, are apparent between the 1909 and 1959 photographs. The window seat to the left of the fireplace was removed by 1959. The window to the right of the fireplace was covered by panelling, which was removed by 1959. It seems likely that a

[173] Rackham, Harris (ed.) 1939. *Christ's College in former days. Being articles reprinted from the College Magazine*. Cambridge: University Press, p. 267.

"Entrance Court. S. range. Room on first floor, staircase 'G'. Panelling *c.* 1600". *An inventory of the historical monuments in the city of, Cambridge* (1959).

doorway was added leading to the adjoining room in the northwest corner. This was covered with panelling for the 2009 restoration. Also at an unknown date, central heating radiators were added under the two bay window seats.

A formal architectural description of the rooms was published in 1959, coincidentally the centenary of the *Origin of Species.*

> W. of stair 'G' is lined with panelling, said to have come from elsewhere, of c. 1600 and in five heights with frieze-panels carved with scroll enrichment and a dentil-cornice; the frieze and cornice are in part gilded and coloured. In the W. wall are two projecting doorcases, each with fluted pilaster-strips at the sides and a pedimented entablature of unconventional form having a deep frieze carved with a trefoiled shell, flanking foliated brackets supporting the pedimented cornice and a lion's mask in the tympanum; the doors are in six panels. The fireplace is original, with chamfered jambs and moulded four-centred head; it is flanked by modern wood pilasters supporting an overmantel, contemporary with the panelling, comprising four arched panels enriched with guilloche-ornament and divided and flanked by reeded and fluted styles; below the S.E. window are some reused panels with similar arched decoration. The door from this set to stair 'G' is of the late 17th century.[174]

[174] Royal Commission on the Historical Monuments of England. 1959. *An inventory of the historical monuments in the city of Cambridge.* 2 vols. London: Her Majesty's Stationery Office, pp. 34–35.

In 2008 the mid-twentieth-century carpeting was removed, exposing nineteenth-century floorboards. The perimeter of the boards was blacked and a 13 × 15 foot section in the centre of the room was bare unstained wood, where a rug would have been. This is visible in the 1909 photograph. This waxing has been preserved. These floorboards are lain perpendicular over an older set of wider boards which are visible just inside the main door and in the bedchamber or cupboard. The under boards in the dressing room and under the radiators (below the window seats) are pine tongue and groove and too narrow and uniform to be original.

In 2008 the College hired Jo Poole, an expert in theatrical and period costume and soft furnishings, to assist with the restoration of the rooms. We worked closely together on research into Cambridge student rooms of the period and spent long evenings in the rooms preparing them. Her involvement led to a major boost in the level of authenticity achieved. I am particularly grateful for her advice and assistance during the restoration.

At Poole's suggestion the College engaged Matthew Beesley, Senior Conservator at Fairhaven Stone Ltd, to analyse the panelling for evidence of former paints from Darwin's time. Tiny fragments were taken from many parts of the room and analysed under the microscope and other forms of analysis. Susan Smith, Conservation and Design Officer at Cambridge City Council, kindly inspected the site with me and after my written application gave permission for the Grade 1 listed property to be painted by Fairhaven according to their analysis. Fairhaven mixed paints for the rooms that are most likely to be the correct colours for Darwin's time. The rooms were painted between October 2008 and January 2009. A light yellow colour was added to the grooves outside the panels as traces were found that might have shown lightly through the green. More colours were added to the frieze and ornamental doors.

Jo Poole made the most exciting and unexpected discovery in the rooms. When examining the horsehair seat cushions in the bay windows, she found that both cushions were covered in four layers of different seat covers from different periods, each sown over an older layer. She writes: "The outer cover was a cream brocade from the last quarter of the twentieth century. Underneath was a green/brown upholstery fabric from the inter-war period. Below this was a sturdy maroon-coloured cloth from the late nineteenth or early twentieth century which had some basic machine stitching in its construction. Extra padding was stitched below this layer, possibly to stop the itchy horsehairs emerging through the cloth.

Finally, the oldest layer was a blue and beige print on a fairly lightweight cotton fabric. This fabric was directly over the horsehair, and had been stitched entirely by hand with linen thread. These layers were the same on the cushions in both windows."[175] This fabric was carefully recreated and Poole used to sew new seat cushions to cover and protect the original ones and to produce splendid curtains for the rooms. The colours fit the wall colours well.

Poole also commissioned the production of a large woollen carpet like those seen in contemporary drawings from the Enterprise Weaving Company, Kidderminster. The design was one that was available in 1828 and the colours match the fabric found on the seat cushions. Furniture, crockery and many other items were purchased or manufactured to complete the early 19th-century appearance of the rooms. Pete Davenport, an expert in historic firearms, delivered a period percussion shotgun like the one Darwin kept in his rooms. Many people lent items to improve the room display. See the list of items in Appendix 2. In preparation for opening the room to visitors, Darwin's name was added to the list of names painted on the wall at the front of the staircase. In pride of place was Darwin's beetle cabinet lent by Milo Keynes. The results were a complete transformation of the rooms and the closest we can ever come to what Darwin's "most snug & comfortable rooms" must have looked like. At the recommendation of Jo Poole, the photographer Allison Maletz was asked to photograph the room for archival purposes in February 2009. Some of her photographs are below.

The restored rooms were covered in the national and international media, by the arrangement and instruction of the College Development Office, and featured in the magazine *World of Interiors*. Darwin's restored college rooms were opened to the public for the first time in a century. I gave very many tours of the rooms. Some of the most memorable visitors were Sir David Attenborough and on another occasion, a large group of Darwin family and descendants who had gathered for a reunion at Darwin College, to which I had been invited. I suppose never before had so many Darwins gathered there. These rooms are still a working part of the College, used by Fellows for supervisions or tutorials.

---

[175] See Jo Poole, 2009. The restoration of Darwin's rooms. http://www.christs.cam.ac.uk/college-life/article-jo-poole.

Archival photographs of Darwin's restored rooms at Christ's College, February 2009. Courtesy of Allison Maletz.

ACKNOWLEDGEMENTS

It is a pleasure to acknowledge the kind assistance in many ways of the Master and Fellows of Christ's College, Jo Poole, Candace Guite, College Librarian, Colin Higgins, Assistant College Librarian, Christopher Jakes, Principal Librarian Local Studies, Cambridgeshire Libraries, Geoffrey Thorndike Martin, Kees Rookmaaker, Jacqueline Cox of the Cambridge University Archives, David Sedley, David Butterfield, Christopher Woods, Robert Hunt, Richard Keynes, Simon Keynes, Randal Keynes, Susan O'Donnell, Cath Green and Peter Hester. David Norman, Gordon Chancellor, Peter Kjaergaard, Cordula van Wyhe and Christopher Woods provided helpful comments on earlier drafts of this booklet.

Dai Jones, current incumbent of Darwin's rooms and responsible to the Darwin Committee, along with John van Wyhe, and Wayne Bell of the College Maintenance Department, for the restoration of Darwin's rooms, was enormously helpful and encouraging throughout the restoration process and gave me unrestricted use of the rooms during my years at Christ's. Together with Jo Poole and Matthew Beesley and his team at Fairhaven and Woods Ltd, and the staff of the College maintenance department, particularly Wayne Bell, Martin Tuck and Graham Howe, the restoration work was carried out in 2008–9. Other parts of the interior decorations were carried out by J. Halbert Ltd. Jo Poole was greatly assisted by Jane Sargent (fine art conservation), Ali Fletcher and Pete Davenport (antique weapons dealer).

I am grateful to Cambridge Central Library, the Whipple Museum of the History of Science, University of Cambridge and the Syndics of Cambridge University Library and Gordon Chancellor to reproduce Darwin manuscripts and other illustrations from their collections. Other images are from *The Complete Work of Charles Darwin Online* (http://darwin-online.org.uk/) and the collection of the author. Allison Maletz very kindly allowed me to reproduce her beautiful photographs of Darwin's rooms. I am grateful to Frank Kelly, the Master of Christ's College, for permission to re-publish the original portions of this work from 2009. Unless otherwise indicated, all photographs, engravings and other illustrations are by the author or are from the author's collection.

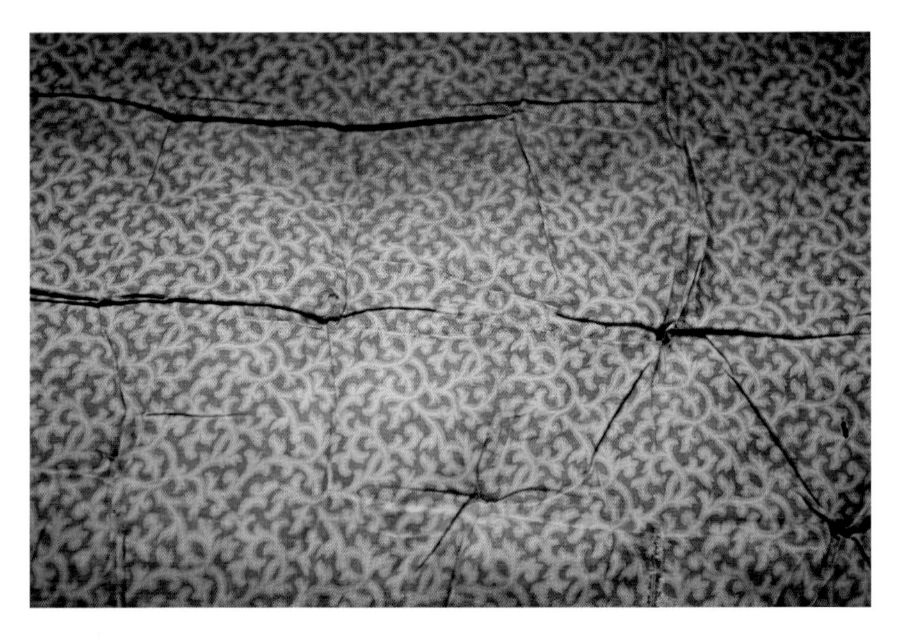

The fabric discovered on the bay window seat cushions.

APPENDIX 1

DARWIN'S COLLEGE BILLS

Until recently the only records of Darwin as a student at Christ's were his name entered in three Admissions Books. The six Record Books below were discovered by Geoffrey Thorndike Martin and are reproduced with the kind permission of the Master and Fellows of Christ's College. These were transcribed by Kees Rookmaaker and first published on *Darwin Online* in 2009. He has reorganized them into condensed tables for presentation here. Students' names are listed according to their seniority as members of the College, determined by when they were enrolled. Darwin was enrolled on 15 October 1827. Students are listed in the Record Books by surname. As Darwin followed his elder brother Erasmus Darwin to Christ's College, the Record Books at first refer to Charles Darwin as 'Darwin junior' and later 'C. Darwin' to distinguish him. Amounts are, of course, written in the pre-decimal system of pounds, shillings and pence, abbreviated as £ s d. A pound was worth 20 shillings and a shilling was 12 pence. Colour images of all of the below pages are available on *Darwin Online*.

Glossary of terms used in the Record Books
ad Babtistam = "those admitted between Ash-Wednesday and St. John Baptist's Day (June 24th)".

Apoth.^{y} = apothecary.
Bedm.^{r} = bed maker, a college domestic servant.
Brazier = a portable heater.
Buttery = the part of the College which issued certain kinds of provisions, especially liquors, bread and butter and other foodstuff served cold. The word does not derive from butter but, like butler, derives from bottle.
Caution money = a deposit paid by an undergraduate to the Tutor against possible future debts.
Chamb: chamber, meaning Darwin's rooms.
Com: Wine / Comb. W. = combination wine.
Commons = College meals in Hall, eaten at a common table with others.
D^{s} = Dominus, the title of all Bachelors of Arts. Students with a Master of Arts degree the title of Mr.
Eve Joiner = unidentified.
L.D.; Ly. Day; L. Day = Lady Day, 25 March, feast of the Annunciation of the Blessed Virgin Mary, the first of the four traditional English quarter days.
Laund: = laundress.
L. Drap. = linen draper.
Mich.^{s} = Michaelmas, 29 September, feast of Saint Michael.
Mids.^{r} = Midsummer Day, 24 June.
W. Drap. = woollen draper.
Porter = College employees who watched the gates, recording entrance of students at night and took letters to and from the post office.
Scull = scullion, kitchen servant.
Sempsst = sempstress = seamstress.
Shoebl.^{r} = shoeblacker, a servant who cleaned and polished shoes.
Shoem.^{r} = shoemaker.
Steward = a college servant who oversaw household management.
Tuit$^{n.}$ = tuition.
Ded.^{s} = deductions.
N.Sums = new sums?

Some of the record books in the archives of Christ's College.

'Tutors' Accounts' 1822–1829 (T.11.26)

This 380 × 560 mm (30 mm thick) volume has a modern binding with a brown spine; most of the front and back covers are red. Some of the 96 folios bear the watermark 'C. Wilmott 1816'. There is no inscription on the spine. On the front cover a black label with printed inscription reads: 'TUTORS ACCOUNTS 1822–1824 [sic]'. The pages are not numbered. The volume starts with accounts of Quarter to L. Day 1822. and ends with Christmas 1829. On the inside front cover, in pencil: 'T.11.26 Tutor's Accounts 1822–1829'. On the first page is written in ink: 'J. Shaw, Christ Coll. Cambridge 1822'.

'Tutors' Accounts' 1822–1829 (T.11.26)

	1827	**1828**			
	Christmas Quarter	**Lady Day Quarter**	**Midsummer Quarter**	**Michaelmas Quarter**	**Christmas Quarter**
	Darwin Junior	Darwin Junior	Darwin Junior	Darwin Junior	C. Darwin
Arrears		3.8.6			
Apothecary					
Barber		0.6.6	1.12.0		0.19.6
Bedmaker			1.1.0	0.10.6	1.1.0
Library Books		16.17.0	3.14.0		
Brazier					0.1.0
Cash				0.1.0 chimney sweep	
Chamber	0.11.0	10.11.0	13.0.0	4.0.0	4.0.0
Coals			2.10.0	1.15.11	2.12.6
Cooks		6.0.0	6.0.0		6.0.0
Glazier					0.12.0
Grocer			1.19.7		2.9.7
Hatter					

(Continued)

(*Continued*)

Eve Joiner			0.2.0 & 0.16.6	0.2.0	3.10.0
Laundress					
Linen Draper					
Woollen Draper			40.5.6 F		
Painter			0.11.0	6.7.11	0.2.0
Porter	0.2.8	0.15.3	1.11.3	0.2.0	13.4.0
Scullion					
Shoeblacker		0.7.0	0.7.0	0.3.6	0.7.0
Shoemaker		2.2.0	4.13.0		
Smith					0.8.6
Steward	0.4.10	7.8.11	13.2.11	0.5.10	8.19.5
Taylor		3.10.7	12.9.3		5.9.5
Tuition	2.10.0	2.10.0	2.10.0	2.10.0	2.10.0
Sums	3.8.6	55.16.9	106.5.0	13.4.10	39.15.3
Deductions					
New Sums	3.8.6	55.16.9	106.5.0	13.4.10	39.15.3
Receipts		55.16.6	106.5.0	13.5.0	39.15.3
Signature		J.S. Apr.22. 1828	J.S. July 1. 1828	J.S.Oct.22. 1828	J.S. March 5. 1829

Christ's College T.11.26 Tutors Accounts

	1829			
	Lady Day Quarter	**Midsummer Quarter**	**Michaelmas Quarter**	**Christmas Quarter**
	C. Darwin	Darwin C.	Darwin C.	Darwin C.
Arrears				15.5.7
Apothecary				
Barber	1.0-6	1.7.6		1.10.0
Bedmaker	1.1.0	1.1.0	0.10.6	1.1.0
Library Books				0.7.6
Brazier		0.1.6		0.2.0
Cash		15.0.0		
Chamber	4.0.0	4.0.0	4.0.0	5.0/0
Coals	1.12.6	3.0.0		3.5.0
Cooks	6.0.0	6.0.0	6.0.0	1.2.10
Glazier	0.4.0			0.8.0
Grocer	2.14.10	5.1.4		1.9.4
Hatter				17.11.6
Eve Joiner		0.11.6		17.11.6
Laundress			0.17.0	1.18.0

(Continued)

(*Continued*)

Linen Draper				0.2.6
Woollen Draper				1.18.3
Painter				
Porter	0.17.1	1.12.4 ½	0.2.9	1.3.0
Scullion	0.4.6	0.4.6		0.4.6
Shoeblacker	0.7.0	0.7.0	0.3.6	0.7.0
Shoemaker				1.8.0
Smith				
Steward	5.0.1	16.3.6	0.5.10	10.9.10
Taylor	1.3.9	1.5.5	0.16.0	14.5.11
Tuition	2.10.0	2.10.0	2.10.0	2.10.0
Sums	26.15.3	58.5.7 ½	15.5.7	74.10.4
Deductions				62.0.0
New Sums	26.15.3	58.5.7 ½		28.0.0
				12.10.4
Receipts	26.15.3	58.5.7		40.10.4
Signature	J.G. May 29. 1829	J.G. July 29		£28 advanced in cash

'Students Bills' 1830–1835 (T.11.27) continues the records of t.11.26, although with printed columns, and also includes seamstress, bricklayer, carpenter and study or room rent.

This large folio volume is in a new binding. The headings and columns are printed. The pages are not individually numbered. The last entry for Darwin is in Midsummer 1830, no later record has been found. This book continues the records of T.11.26 and also includes seamstress, bricklayer, carpenter and study or room rent.

'Students Bills' 1830–1835 (T.11.27)

	1830				1831	
	Lady Day Quarter	**Midsummer Quarter**	**Michaelmas Quarter**	**Christmas Quarter**	**Lady Day Quarter**	**Midsummer Quarter**
	C. Darwin	C. Darwin	C. Darwin	C. Darwin	Darwin	Darwin
Balance last Qr			53.9.11	13.17.4	77.13.9	
Bedmaker	1.1.0	1.1.0	0.10.6	1.1.0	1.1.0	1.1.0
Cash	0.6.8	5.0.0			5.3.10 ½	
Coals	4.12.6	2.0.0		3.17.6	3.12.6	2.0.0
Cook	6.0.0	6.0.0	6.0.0	6.0.0	6.0.0	6.0.0
Laundress		6.1.1			2.18.10 ½	3.1.6 ½
Porter	1.4.10	1.3.4		1.8.1	1.7.5	1.0.2
Scullion	0.4.6		0.5.0	0.2.0		0.1.0
Seamstress						0.1.6
Shoeblack	0.7.0	0.7.0	0.3.6	0.7.0	0.7.0	0.7.0
Steward	15.18.2	7.9.2	0.5.10	13.9.7	10.11.1	8.12.4
Study Rent	4.0.0	5.0.0	4.0.0	4.0.0	4.0.0	4.0.0
Tutor	2.10.0	2.10.0	2.10.0	2.10.0	2.10.0	2.10.0

(Continued)

(Continued)

College Account	36.4.8	22.19.7	67.4.9	46.12.6	114.18.6	28.14.6 ½
Apothecary						
Barber	2.1.0	0.3.6		1.5.0	1.9.6	1.10.0
Bookseller						
Circ. Library						
Brazier				0.5.1		0.2.0
Chimney	0.1.0					
.Sweeper						
Carpenter	0.1.6	0.3.0	0.2.6		-.5.6	
Glazier	0.3.6	0.2.6		0.5.9		0.2.
Grocer	5.16.4	1.6.0		5.1.11	3.17.9	1.17.3
Hatter						
Linen draper		3.8.9				
Painter	0.3.9			0.2.0		
Shoemaker	1.16.0	2.1.0		2.16.6	1.14.6	1.1.4.6
Smith						
Tailor	3.1.5	1.12.3		0.5.0		0.12.3
Upholsterer						
Private Tuition				21.0.0		

(Continued)

(*Continued*)

Tradesmen's Bills	13.4.6	8.17.-	0.2.6	31.1.3	7.7.3	4.18.0
Total	49.9.2	31.16.7	67.7.3	77.13.9	122.5.9	33.12.6 ½
Scholarships		17.8.0				
Balance last Qr			53.9.11			
Balance Dr.		14.8.7	13.17.4	77.13.9		33.12.6 ½ [rubbed out?]
Received	49.10.0				122.5.9	
Date	June 4, 1830				May 1831	

'Students Bills' 1827–1831 (T.11.25) records Darwin's weekly meal account in college — and a fragment of a medical excuse note.

This record book records Darwin's weekly commons, buttery account and similar expenses in College. The book is bound in dark brown vellum. Entries are written towards the spine on the verso page and continued from the spine on the recto page opposite. The financial quarter is recorded in the upper left corner. Names of students are in the far left corner.

Lady Day Quarter — 1828 — begins 14 December 1827

Week no.	Darwin	
1		
2		
3		
4		
5		
6	in 26 January 1828	
7	0.11.3	3 and 4
8	0.9.5	3 and 4
9	0.8.0	3 and 4
10	0.7.6	3 and 4
11	0.7.3	3 and 4
12	0.6.2¼	1 and 4
13	0.6.9	
Buttery account		2.6.8
In & out of commons		In 26 January 1828
No. of Days		49
Price of Commons		3.14.4 ½
Vegetables &c at 5 ½*d.* per day		1.2.5 ½
Various posts, total		0.4.10
Names writing		0.06
Total		7.8.11

(*Continued*)

(*Continued*)

Midsummer Quarter — 1828 — begins 14 March 1828

1	**Darwin**	
2	0.9.1	3 and 4
3	0.5.9	1 and 4
4	0.7.8	1 and 4
5	0.7.9	
6	0.9.8	1 and 4
7	0.10.5	3 and 4
8	0.14.11	3 and 4
9	0.12.6	1 and 4
10	0.17.7	3 and 4
11	0.9.1	3 and 4
12	013.1	
13		
Buttery account		5.17.10
In & out of commons		In 15 March 1828,
No. of Days		74
Price of Commons		5.6.2 ½
Vegetables &c at 5 ½*d.* per day		1.13.11
Various posts, total		0.4.10
Names writing		
Total		13.2.11

Michaelmas Quarter 1828 — begins 13 June 1828

Week no.	**Darwin**
1	
2	
3	
4	
5	

(*Continued*)

(*Continued*)

6	
7	
8	
9	
10	
11	
12	
13	
Buttery account	
In & out of commons	
No. of Days	
Price of Commons	
Vegetables &c at 5 ½*d.* per day	
Various posts, total	0.5.4
Names writing	0.0.6
Total	0.5.10

Christmas Quarter 1828 — begins 12 September 1828

Week no.	Darwin	
1		
2		
3		
4		
5		
6		
7	In 31 October 1828	
8	0.12.7	3 and 4
9	0.14.3	1 and 4
10	0.14.11	3 and 4
11	0.5.6	1 and 4

(*Continued*)

(*Continued*)

12	0.12.2	1 and 4
13	0.15.6 and 0.14.8	1 and 4
Buttery account		4.19.12
In & out of commons		In 1 Nov & 9 Nov
No. of Days		30
Price of Commons		2.16.0
Vegetables &c at 5 ½*d.* per day		0.17.10 ½
Various posts, total		0.4.10
Names writing		0.0.10
Total		8.19.5

Lady Day Quarter 1829 — begins 12 December 1828

Week no.	Darwin	
1		
2		
3		
4		
5		
6		
7	In 24 February	
8		
9		
10	0.11.2	3 and 4
11	0.25.10	3 and 4
12	0.31.11	3 and 4
13	0.30.11	3 and 4
In & out of commons		
Total		5.0.0 ½

Midsummer Quarter — 1829 — begins 20 March 1829

Week no.	Darwin	
1	1.7.5	3 and 4
2	1.3.5	2
3	1.14.2	4
4	1.5.9	4
5	1.4.6	4
6	o.12.7	4
7	0.18.1	3
8	1.7.11	4
9	1.5.7	4
10	1.10.5	
11	2.1.10	3
12	1.6.0	3
13	0.4.10 ½	
In & out of commons		Out 8 June
Total		16.3.5 ½

Michaelmas Quarter — 1829 — begins 20 June 1829

Week no.	Darwin	
1		
2		
3		
4		
5		
6		
7		
8		
9		
10		
11		
12		
13		
In & out of commons		
Total		0.5.10

Christmas Quarter — 1829 — begins 25 September 1829

Week no.		
1		
2		
3	0.13.8	4
4	1.7.9	4
5	1.8.9	4
6	1.5.10	3
7	1.11.9	4
8	1.4.4	
9	1.17.8	4
10	1.8.4	3
11	1.6.7	4
12		
13	0.5.4	2
In & out of commons		12 October
Total		12.3.0

Lady Day Quarter — 1830 — begins 25 December 1829

Week no.	Darwin	
1		
2	1.7.11	4
3	1.9.9	
4	1.3.9	3
5	1.3.8	2
6	1.3.6	2
7	1.8.10	4
8	1.4.2	4
9	1.5.0	3
10	1.6.8	1

(*Continued*)

(*Continued*)

11	1.8.8	4
12	1.4.6	
13	1.11.4	4
In & out of commons		in 1 Jan
Total		15.18.1

Midsummer Quarter — 1830 — begins 25 March 1830

Week no.	Darwin	
1	1.1.11	3
2	1.7.6	
3	1.2.8	4
4	1.2.10	4
5	1.1.4	3
6	1.5.8	3
7	1.7.6	4
8	1.0.5	4
9	1.8.1	4
10	1.5.7	4
11	0.7.11	2
12		
13	0.9.10	
In & out of commons		out 3 June
Total		12.11.7

Michaelmas Quarter — 1830 — begins 25 June 1830

Week no.	Darwin
1	
2	
3	

(*Continued*)

(*Continued*)

4		
5		
6		
7		
8		
9		
10		
11		
12		
13		
In & out of commons		
Total		0.5.10

Lady Day Quarter — 1831 — begins 24 December 1830

Week no.	Darwin	
1	0.18.4	2
2	0.18.4	2
3	0.18.9	2
4	0.18.8	2
5	1.2.9	
6	0.18.4	2
7	1.0.7	
8	1.0.4	
9	0.19.8	2
10	0.17.9	2
11		
12		
13	0.10.4	
In & out of commons		
Total		10.4.1

Midsummer Quarter — 1831 — begins 24 March 1831

Week no.	Darwin	
1		
2		
3		
4	0.1.5	2
5	1.3.2	2
6	1.1.1	2
7	1.4.11	2
8	1.6.4	2
9	1.2.5	2
10	1.3.0	
11	1.19.8	
12		
13		
In & out of commons		B.A. 25 April
Total		8.12.3 ½

Michaelmas Quarter — 1831 — begins 11 June 1831

Week no.	Darwin
1	1.8.6 ½
2	
3	
4	
5	
6	
7	
8	
9	

(*Continued*)

(*Continued*)

10		
11		
12		
13	0.10.10	
In & out of commons		out 16 June
Total		1.19.4 ½

Christmas Quarter — 1831 — begins 23 September 1831

Week no.	Darwin	
1		
2		
3		
4		
5		
6		
7		
8		
9		
10		
11		
12		
13		
In & out of commons		
Total		0.9.10

Lady Day Quarter — 1832 — begins 23 December 1831

Week no.	Darwin
1	
2	
3	

(*Continued*)

(*Continued*)

4	
5	
6	
7	
8	
9	
10	
11	
12	
13	
In & out of commons	
Total	0.10.4

'Study Rent' 1828–1831 (T.9.5)

This volume of records quarterly room rent for October 1828 — June 1849. The book is bound in brown vellum. The pages are unnumbered.

> Copy of an agreement at a Coll. meeting April 20, 1831.
>
> Agreed that the Rent of the Ground Floor Rooms & the Rooms over the Kitchen in the new building be Fifteen Pounds, & of the other Rooms Twenty.
>
> And that the rent of all the rooms in the old buildings occupied by Fellow Commoners be Twenty Pounds, & of the rooms on the South side of the old Court occupied by Pensioners be Fifteen, & of those on the North side be Twelve Pounds.
>
> Garretts in the Front Court to pay a rent of Four pounds.
>
> Mem. An additional charge is made of 10s/ a Quarter on each Fellow Commoner, 5s on each resident Bachelor, & 5s on all the Pensioners and Ten-year men, in aid of the new Building.

Year and Quarter	Name	Amount
1828 Christmas Quarter	C. Darwin	4.0.0
1829 Lady Day Quarter	Darwin C.	4.0.0
1829 Midsummer Quarter	Darwin C.	4.0.0
1829 Michaelmas Quarter	Darwin C.	4.0.0
1829 Christmas Quarter	C. Darwin	4.0.0
1830 Lady Day Quarter	Darwin C.	5.0.0
1830 Midsummer Quarter	Darwin C.	4.0.0
1830 Michaelmas Quarter	Darwin C.	4.0.0
1830 Christmas Quarter	Darwin	4.0.0
1831 Lady Day Quarter	Darwin	4.0.0
1831 Midsummer Quarter	Darwin	4.0.0
1831 Michaelmas Quarter	[Darwin among many not listed]	
1831 Christmas Quarter	[Darwin among many not listed]	
1832 Lady Day Quarter	C. Darwin	[blank]
1832 Midsummer Quarter	[Darwin among many not listed]	
1832 Michaelmas Quarter	Darwin	[blank]
1832 Christmas Quarter	[Darwin among many not listed]	

1828–1829 'Residents Book' (T.17.A)

This book records weekly residence in College and terms kept. This book has a modern binding.

Midsummer Quarter — 1828 — begins in March 15

Week no.	Saturday	Darwin
1	15 March 1828	4 and 3 days
2	22 March 1828	4 and 3 days
3	29 March 1828	4 and 3 days
4	5 April 1828	4 and 3 days
5	12 April 1828	4 and 3 days

(*Continued*)

(*Continued*)

6	19 April 1828	4 and 3 days
7	26 April 1828	4 and 3 days
8	7 May 1828	4 and 3 days
9	10 May 1828	4 and 3 days
10	17 May 1828	4 and 3 days
11	24 May 1828	4 days
12	31 May 1828	
13	7 June 1828	
Total		11 weeks, 74 days
1*s.* 6*d.* per day = 5.4.10		
1½*d.* per week = 0.1.4½		
Total charge 5.6.2½		

Michaelmas Quarter — 1828 — no entries for Darwin in the ledger.
Christmas Quarter — 1828 — begins 12 September

Week no.	Saturday	Darwin
1	13 September 1828	4 and 3 days
2	20 September 1828	
3	27 September 1828	
4	4 October 1828	
5	11 October 1828	
6	18 October 1828	
7	25 October 1828	
8	1 November 1828	4 days
9	8 November 1828	
10	15 November 1828	4 and 3 days
11	22 November 1828	4 and 3 days
12	29 November 1828	4 and 3 days

(*Continued*)

(*Continued*)

13	6 December 1828	4 and 3 days
Total		6 weeks, 39 days
1*s*. 6*d*. per day = 2.15.3		
1½*d*. per week = 0.0.9		
Total charge 2.16.0		

Lady Day Quarter — 1829 —

Week no.	Saturday	Darwin
1		
2		
3		
4		
5		
6		
7		
8		
9		
10		3 days
11		4 and 3 days
12		4 and 3 days
13		4 and 3 days
Total		
No calculations		

Midsummer Quarter — 1829 — begins 20 March

Week no.	Saturday	Darwin
1		4 and 3 days
2		4 and 3 days
3		4 and 3 days

(*Continued*)

(*Continued*)

4	4 and 3 days
5	4 days
6	4 days
7	4 days
8	4 and 3 days
9	4 and 3 days
10	4 and 3 days
11	4 and 3 days
12	4 days
13	
Total	
[No calculation]	

Michaelmas Quarter — 1829 — begins 19 June 1829

No entries for Darwin.

'Lecturer's Book' 1831–1836 (T.8.2)

1831 —
Bachelor's Commencement
D^{s}. Darwin B.A } ad Baptistam — 14.0.0
1836.
Master of Arts Commencement.
Darwin. M. A } — 12.0.0

APPENDIX 2

List of Contents on Display in Charles Darwin's Rooms at Christ's College. [2009]

By Dr John van Wyhe, Bye-Fellow

Panelling Sixteenth-century panelling originally from other buildings, now painted as it appeared in Darwin's day. The two ornamental doorways in the west wall open onto a brick wall.	Christ's College (CC)
Curtains Made by Jo Poole in accordance with contemporary techniques from cotton fabric replicated from the original seat cushion covers found in the room.	CC

Carpet	CC
Brussels carpet (looped wool pile) made by the Enterprise Weaving Company, Kidderminster, from a design available in 1828.	
Desk	CC
Early nineteenth-century knee-hole desk	
Writing box with two ink wells with silver lids, quill pen, ivory/silver pencil and silver/mother of pearl penknife.	Lent by Penny Price-Larkum
Reproduction of Darwin letter to W. D. Fox [5 April 1830]	Original letter in CC Old Library.
Five British bird skins: barn owl, kestrel, water rail, little bittern, corncrake	Lent by La Societé Sercquaise, Isle of Sark, courtesy of Richard Axton
Three local fossils (from Coton)	Lent by Penny Price-Larkum
Sample of pumice stone	Lent by John van Wyhe
Pill boxes	Lent by John van Wyhe
Stephens, *British entomology: Mandibulata* vol. 1 (on bookstand)	CC Old Library
Oak bookstand	Lent by John van Wyhe
Wooden spool of grey thread	Lent by John van Wyhe
Oil lamp converted to electric	CC
Dining table	CC
Georgian drop leaf table	
Damask table cloth	CC
Four Regency dining chairs	Lent by Candace Guite
Christ's College pensioners gown (draped over dining chair) and cap	CC
Two silver candle holders with glass chimneys and silver snuffers.	Lent by Penny Price-Larkum
Silver salt "shell" with spoon	Lent by Penny Price-Larkum
Damask napkins	Lent by Penny Price-Larkum
Contemporary plates and small ale glasses	CC

Contemporary wine bottle	CC
Contemporary playing cards	CC
Bedroom	
Bed with contemporary quilt	CC
Wash stand, early nineteenth century.	CC
Brass petal-shaped candle holder with beeswax candle	Lent by Penny Price-Larkum
Oil lamp converted to electric	CC
Jug and bowl, chamber pot, ceramic hot water bottle	CC
On the walls	
(Above desk) Reproduction of an engraving of Leonardo da Vinci by Raphael Morghen. A copy at Down House bears the inscription "Gift of Leonard Darwin, 1929" and "This belonged to Father".	CC
(Left of bookcase) Reproduction of engraving of Raphael Sanzio's Madonna della Sedia by Desnoyers. A copy of this engraving was owned by E. A. Darwin.	CC, Reproduced courtesy of the Fitzwilliam Museum, Cambridge, acc. no. 33.A.1-107.
(Right of bookcase) Reproduction of an engraving by Marcantonio Raimondi of Raphael Sanzio's Apollo on Parnassus, a mural in the Vatican. Possibly one of the engravings Darwin displayed in his rooms at Christ's.	CC, Reproduced courtesy of the Fitzwilliam Museum, Cambridge, acc. no. P.5323-R.
Wedgwood medallion portrait of Charles Darwin on green jasper by Thomas Woolner. Second panel reads "CHARLES ROBERT DARWIN 1829-31". The date 1829 is a mistake for 1828. "Erected by G. H. Darwin Plumian Professor 1885".	CC
Reproduction armchair (below medallion)	Lent by John van Wyhe

Gun rack

English 1820s double-barrelled percussion shotgun	CC

"In the latter part of my school life I became passionately fond of shooting; I do not believe that any one could have shown more zeal for the most holy cause than I did for shooting birds. How well I remember killing my first snipe, and my excitement was so great that I had much difficulty in reloading my gun from the trembling of my hands. This taste long continued, and I became a very good shot. When at Cambridge I used to practise throwing up my gun to my shoulder before a looking-glass to see that I threw it up straight. Another and better plan was to get a friend to wave about a lighted candle, and then to fire at it with a cap on the nipple, and if the aim was accurate the little puff of air would blow out the candle. The explosion of the cap caused a sharp crack, and I was told that the tutor of the College remarked, 'What an extraordinary thing it is, Mr Darwin seems to spend hours in cracking a horse-whip in his room, for I often hear the crack when I pass under his windows.'" Darwin, *Autobiography*.

Wood and brass cleaning rod	CC
White linen shot pouch	CC
Leather and brass shot flask with adjustable charger	CC
Brass powder flask	CC
Corked powder jug	CC
Percussion cap tin	CC
Bottle of gun oil	CC
Leather satchel	Lent by John van Wyhe
Brass candle holder	CC
Wadding punch, wadding and wool rags and leather wadding pouch	CC
Knotted string	

"I kept an exact record of every bird which I shot throughout the whole season... which I used to do by making a knot in a piece of string tied to a button-hole." Darwin, *Autobiography*

Shooting stick	CC
Mirror (Regency Pier glass)	CC
Drop leaf/ Pembroke table	Lent by Candace Guite
Mahogany beetle collecting cabinet with cork-lined drawers (possibly Darwin's own cabinet made for this room)	Lent by Milo Keynes, great-grandson of Charles Darwin.
Collection of British beetles in beetle cabinet	Lent by Christopher Lewis
1830s red leather notebook	Lent by John van Wyhe
Geological hammer	Lent by John van Wyhe
Corked collecting bottle with insects	Lent by John van Wyhe
Insect sweeping net	Lent by Ian Ferguson
Pill boxes	Lent by John van Wyhe
Brass tweezers (originally included with the microscope)	CC
Pocket lens in leather cover	Lent by John van Wyhe
Foot stool	CC, Old Library
Table by fireplace	CC
Early nineteenth-century Pembroke table	
Regency teapot	Lent by Elisabeth Everitt
Contemporary tea cups, coffee cans, saucers, spoons, slop bowl, milk jug	CC
Oil lamp converted to electric	CC
Mantelpiece	
Two hand-thrown terracotta flower pots with moss and moth chrysalis	Lent by John van Wyhe, moss courtesy of Jo Poole, chrysalis courtesy Martin Tuck

"The Chrysalis goes on very well, it is much more lively, so that if touched, it will roll itself about. — & it appears to me the parts about the head are very much more distinct than they formerly were. — The process is, — a flower pot-full of dampish mold, & over that a stratum of not very dry sand, on which the Chrysalis is placed; an inverted glass vessel, which I suppose prevents too much evaperation, is placed over it, — the whole kept in a warm room." Darwin to W. D. Fox, [7 Jan. 1829], *Correspondence*, vol. 1, p. 73.

Contemporary tankard	CC
Clay tobacco pipe	Lent by John van Wyhe

West bay window

Original horse hair seat cushion and cotton cover	CC
Gould-type microscope mounted on a velvet lined mahogany box with three objectives, stage forceps, a pair of multi-celled bone sliders, bone handled dissecting needle, small glass and sealing wax trough and a glass test slide.	CC

Darwin was given a Gould-type microscope by J. M. Herbert in early May 1831 with the accompanying note: "If M[r]. Darwin will accept the accompanying Coddington's Microscope, it will give peculiar gratification to one who has long doubted whether M[r]. Darwin's talents or his sincerity be the more worthy of admiration, and who hopes that the instrument may in some measure facilitate those researches which he has hitherto so fondly and so successfully prosecuted." The original microscope is now at Down House.

East bay window

Birds nest coated with minerals under glass dome.	Lent by the Sedgwick Museum of Earth Sciences, Cambridge

"Professor Sedgwick happened one day to mention a spring issuing from one of the chalk hills at Trumpington or Coton which deposited carb. of lime very prettily upon twigs &c. Darwin said to me, "I shall go and test that water for myself", which he did and found the fact to be as Sedgwick had stated it. Not content with this he deposited a large bush in the spring and at a subsequent lecture presented it to Sedgwick who exhibited it as being, what it really was, a very beautiful specimen. Several members of Sedgwick's class followed D's example and adorned their rooms with similar specimens of Increstation." Cambridge contemporary J. M. Rodwell, CUL-DAR112.B118-B121, transcribed on *Darwin Online*.

Original horse hair seat cushion and cover; protected by new cover by Jo Poole.	CC

Reproduction George III bookcase	**CC**
Milton, *Paradise lost*. London, 1764.	Lent by John van Wyhe

Anon, [Greek New Testament] Leipzig, 1850. CC Old Library

Dillwyn, *British Confervae; or colored figures and descriptions of the British plants referred by botanists to the genus conferva*. London, 1809. [Copy owned by John Stevens Henslow, bearing his bookplate inside front cover.] CC Old Library

Herschel, *A preliminary discourse on the study of natural philosophy*. London, 1831. CC Old Library

Humboldt, *Personal narrative of travels to the equinoctial regions of the new continent, during the years 1799–1804*. vol. 5 London, 1821. CC Old Library

Kirby and Spence, *An introduction to entomology*. 4 vols. London, 1822. (Vol. 4 is next to the beetle cabinet) CC Old Library

Paley, *Natural theology*. vol. 2 Oxford, 1826. CC Old Library

Rose, *Eight sermons preached before the university of Cambridge at Great St. Mary's, in the years 1830 and 1831*. Cambridge, 1831. CC Old Library

Ross, *Hirsch's geometry; or a sequel to Euclid*. London, 1827. CC Old Library

Samouelle, *The entomologist's useful companion; or an introduction to the knowledge of British insects*. London, 1819. CC Old Library

Simson, *The elements of Euclid*. Glasgow, 1817. CC Old Library

Smith, *The English Flora*. vols. 1-4, 2^d^ edn. London, 1828. CC Old Library

Stephens, *Illustrations of British entomology*. London 1821–1831. CC Old Library

Fireplace

Regency hob grate	CC
Brass fender	CC
Cast iron kettle (Georgian)	CC
Contemporary tongs, pokers etc.	CC
Bellows	Lent by John van Wyhe
Toasting fork (hanging to left of fireplace)	CC
Dog basket	CC

Darwin kept a shooting dog named Dash while a student at Christ's. For a time he also kept a dog named Sappho left by W. D. Fox.

Two reproduction leather easy chairs in the style of George III	CC

REFERENCES

Alpha Beta. 1803. An account of Cambridge. *Monthly Magazine and British Register,* 15 (February, March): 26–31, 117–120.

Anon. 1824. *Gradus ad Cantabrigiam: or, new university guide to the academical customs, and colloquial or cant terms peculiar to the University of Cambridge; observing wherein it differs from Oxford.* London: J. Hearne.

Anon. 1829. *The Cambridge University calendar for the year 1829.* Cambridge: J. Deighton.

Anon. 1829. *The Lion* 3 (21): 641.

Anon. 1830. *Classical examinations; or, a selection of University scholarship and other public examination papers, and of the question papers on the lecture subjects of the different Colleges in the University of Cambridge.* Cambridge: Grant.

Anon. 1830. *The Cambridge guide, or a description of the University and town of Cambridge.* Cambridge: Deighton.

Anon. 1835. [Report of a meeting of the Cambridge Philosophical Society]. *The Times* (22 December): 7.

Anon. 1836–7. [Reports of Darwin's communications read to the Cambridge Philosophical Society 1835–1837]. *London and Edinburgh Philosophical Magazine and Journal of Science* 8 (43, January 1836): 79, 80; 10 (61, April 1837): 316.

Anon. 1840. *The costumes of the members of the University of Cambridge.* London: N. Whittock.

Anon. 1862. *Student's guide to the University of Cambridge.* Cambridge: Deighton.

Anon. 1879. University intelligence [meeting in Christ's College, Cambridge of the committee of the Darwin Memorial Fund]. *The Times* (8 March): 9.

Anon. 1881. The scientific education of woman [with a donation by Darwin for a Cambridge laboratory for women]. *The Times* (27 January): 4.

Anon. 1900. List of donations [books] received during the year 1899: From the executors of the late Mrs Darwin. *Cambridge University Reporter* 30 41. 15 June: 1079–1080.

Anon. 1909. Darwin's lodgings. *Christ's College Magazine* (Easter Term): 248–250.

[Atkinson, S.] 1825. Struggles of a poor student through Cambridge. *London Magazine and Review* (new series) 1: 491–510.

Attenborough, David. 2009. *Life stories*. London: HarperCollins.

Autobiography: 1958. *The autobiography of Charles Darwin 1809–1882. With the original omissions restored. Edited and with appendix and notes by his grand-daughter Nora Barlow*. London: Collins.

Barlow, Nora (ed.) 1967. *Darwin and Henslow. The growth of an idea*. London: Bentham-Moxon Trust, John Murray.

Barrett, Paul H. *et al.* 1987. 1960. *Handlist of Darwin papers at the University Library Cambridge*. Cambridge: University Library.

Barrett, Paul H. 1974. The Sedgwick-Darwin Geologic Tour of North Wales. *Proceedings of the American Philosophical Society* 118: 146–164.

Birds: Gould, John. 1838–1841. *Part 3 of The zoology of the voyage of H.M.S. Beagle. Birds. By John Gould. Edited and superintended by Charles Darwin*. London: Smith Elder and Co.

Blomefield, L. 1887. *Chapters in my life*. Bath: [privately printed].

Browne, Janet. 1995. *Charles Darwin*, vol. 1: *Voyaging*. London: Pimlico.

Bury, J. P. T. 1967. *Romilly's Cambridge diary 1832–1842*. Cambridge: University Press.

Butler, Thomas. 1882. [Recollections of Darwin, 13 September] (DAR 112.A10-A12), transcribed by Kees Rookmaaker. *The Complete Work of Charles Darwin Online* (http://darwin-online.org.uk/).

Carlisle, N. 1818. *A concise description of the endowed Grammar Schools in England*, vol. 2. London: Baldwin, Cradock and Joy.

Chancellor, Gordon. 2012 Humboldt's Personal narrative and its influence on Darwin. Essay available on *The Complete Work of Charles Darwin Online* (http://darwin-online.org.uk/EditorialIntroductions/Chancellor_Humboldt.html).

Chancellor, Gordon and Wyhe, John van, with the assistance of Kees Rookmaaker (eds). 2009. *Charles Darwin's notebooks from the voyage of the Beagle*. Cambridge: University Press.

Clark, J. W. and Hughes, T.M. (eds.) 1890. The walking tour in North Wales. In *The life and letters of the Reverend Adam Sedgwick*, vol. 1: 379–381.

Combe, William. 1815. *A history of the University of Cambridge: its colleges, halls, and public buildings*. London: R. Ackermann.

Correspondence: Burkhardt *et al.* eds. 1985–. *The correspondence of Charles Darwin*. 20 vols. Cambridge: University Press.

Corsi, P. 1998: A Devil's Chaplain's calling? *Journal of Victorian Culture* 3 (1): 129–137.

Coyne, Jerry A. 2009. *Why evolution is true*. Oxford: Oxford University Press.

Darwin Manuscript Catalogue on Darwin Online, http://darwin-online.org.uk/MScatintro.html.

Darwin, C. R. 1835. *[Extracts from letters addressed to Professor Henslow]*. Cambridge: [privately printed].

Darwin, C. R. 1827. [Edinburgh notebook]. (DAR 118), transcribed by Kees Rookmaaker. *The Complete Work of Charles Darwin Online* (http://darwin-online.org.uk/).

Darwin, C.R [c. 1827.] [Notes on reading Sumner's Evidence of Christianity]. (CUL-DAR91.114-118), transcribed by John van Wyhe, *The Complete Work of Charles Darwin Online* (http://darwin-online.org.uk/).

Darwin, C. R. 1836. Geological notes made during a survey of the east and west coasts of S. America, in the years 1832, 1833, 1834 and 1835, with an account of a transverse section of the Cordilleras of the Andes between Valparaiso and Mendoza. [Read 18 November 1835] *Proceedings of the Geological Society* 2: 210–212.

Darwin, C. R. 1838. 'Work finished If not marry' [Memorandum on marriage]. (DAR210.8.1), transcribed by Kees Rookmaaker. *The Complete Work of Charles Darwin Online*. Darwin Online, http://darwin-online.org.uk/content/frameset?keywords=finished%20work&pageseq=2&itemID=CUL-DAR210.8.1&viewtype=side.

Darwin, C. R. 1839. *Journal of researches into the geology and natural history of the various countries visited by H.M.S. Beagle etc.* London: Henry Colburn.

Darwin, C. R. 1845. *Journal of researches into the natural history and geology of the countries visited during the voyage of H.M.S. Beagle round the world, under the Command of Capt. Fitz Roy, R.N.* 2nd edition. London: John Murray.

Darwin, C. R. 1859. *On the origin of species by means of natural selection, or the preservation of favoured races in the struggle for life*. London: John Murray.

Darwin, C. R. 1862. [Recollections of Professor Henslow]. In Jenyns, L., *Memoir of the Rev. John Stevens Henslow M.A., F.L.S., F.G.S., F.C.P.S., late Rector of Hitcham and Professor of Botany in the University of Cambridge*. London: Van Voorst, pp. 51–55.

Darwin, C. R. 1868. *The variation of animals and plants under domestication*. 2 vols. London: John Murray.

Darwin, C. R. 1871. *The descent of man, and selection in relation to sex*. 2 vols. London: John Murray.

Darwin, C. R. 1872. *The expression of the emotions in man and animals*. London: John Murray.

Darwin, C. R. 1876–1882. Recollections of the development of my mind & character [*Autobiography*, CUL-DAR26.1-121), transcribed by Kees Rookmaaker. *The Complete Work of Charles Darwin Online* (http://darwin-online.org.uk/).

Darwin, C. R. *et al.* 1878. [Memorial to the Vice-Chancellor respecting the Examination in Greek in the Previous Examination]. *Cambridge University Reporter* (7 December): 206–207.

Darwin, Francis. (ed.) 1887. *The life and letters of Charles Darwin.* 3 vols. London: Murray.

Darwin, Francis. (ed.). 1909. Some letters from Charles Darwin to Alfred Russel Wallace. Darwin Centenary Number. *Christ's College Magazine* 23 (Easter Term): 229.

Darwin, Francis. 1914. William Erasmus Darwin, 1839–1914. *Christ's College Magazine* 29: 16–23.

Darwin, G. and Darwin, F. (eds.) 1909. *Darwin celebration, Cambridge, June, 1909. Speeches delivered at the banquet held on June 23rd.* Cambridge: Cambridge Daily News.

Desmond, Adrian and Moore, James. 1991. *Darwin.* London: Michael Joseph.

Forster, L. M. 1883. [Recollections of Darwin, January.] (DAR 112.A31-A37), transcribed by Kees Rookmaaker. *The Complete Work of Charles Darwin Online* (http://darwin-online.org.uk/).

Fox, W. D. 1824–1826. [Diary and accounts 'No VI' 'Cambridge'] Cambridge University Library, DAR 250.5. Available on *The Complete Work of Charles Darwin Online* (http://darwin-online.org.uk/content/frameset?pageseq=1&itemID=CUL-DAR250.5&viewtype=image).

Fyfe, A. 1997. The reception of William Paley's natural theology in the University of Cambridge. *British Journal for the History of Science* 30: 321–335.

Garland, M.M. 1980. *Cambridge before Darwin: the ideal of a liberal education*, 1800-1860. Cambridge: University Press.

Geikie, A. 1909. *Charles Darwin as geologist: The Rede Lecture given at the Darwin Centennial Commemoration on 24 June 1909.* Cambridge: University Press.

Grant, R.E.. 1827. Notice regarding the ova of the Pontobdella muricata, Lam. *Edinburgh Journal of Science* 7 (July): 160–161.

Harmer, S. F. 1901. List of specimens collected on the *Beagle* which were kept or discarded, with extracts from Darwin's manuscripts referring to specimens kept in the museum. University Museum of Zoology, Cambridge. Image (UMZC-Histories 4.945).

Harraden, Richard. 1805. *Costume of the various orders in the University of Cambridge.* Cambridge: R. Harraden.

Herbert, J. M. 1882. [Recollections of Darwin at Cambridge, 2 June]. (DAR 112. B57-B76), transcribed by Kees Rookmaaker. *The Complete Work of Charles Darwin Online* (http://darwin-online.org.uk/).

Herbert, J. M. 1882. [Recollections of Darwin, 12 June.] (DAR 112.A60-A61), transcribed by Kees Rookmaaker. *The Complete Work of Charles Darwin Online* (http://darwin-online.org.uk/).

Herbert, Sandra. 2002. Charles Darwin's notes on his 1831 geological map of Shrewsbury. *Archives of Natural History* 29 (1): 27–30.

Herbert, Sandra. 2005. *Charles Darwin, geologist.* Ithaca, N.Y: Cornell University Press.

Herschel, J. F. W. 1831. *Preliminary discourse on the study of natural philosophy.* London: Longmans, Rees, Orme, Brown, Green and Longman.

Herschel, J. F. W. 1833. *A treatise on astronomy.* In Lardner, D. (ed.) *The cabinet cyclopædia.* London: Longman, Orme, Brown, Green & Longmans, and John Taylor.

Hitchens, Christopher., *Arguably: essays by Christopher Hitchens.* 2011.

Humboldt, Alexander von and Bonpland, Aimé. 1814–29. *Personal narrative of travels to the equinoctial regions of the new continent, during the years 1799–1804. ... Translated into English by Helen Maria Williams.* 2nd ed. 7 vols. London: Longman, Hurst, Rees, Orme and Brown.

Humphreys, H. N. and Westwood. J. O. 1845. *British moths and their transformations*, vol. 1. London: William Smith.

Jenyns, Leonard.1842. *Fish.* Part 4 of *The zoology of the voyage of HMS Beagle.* Edited and superintended by Charles Darwin. London: Smith Elder and Co.

Jenyns, Leonard, 1862. *Memoir of the Rev. John Stevens Henslow M.A., F.L.S., F.G.S., F.C.P.S., late Rector of Hitcham and Professor of Botany in the University of Cambridge.* London: Van Voorst.

Journal: Wyhe, J. van (ed.) 2005. Darwin's personal 'Journal' (1809–1881). (DAR 158), transcribed by the *Correspondence of Charles Darwin* project. *The Complete Work of Charles Darwin Online* (http://darwin-online.org.uk/).

Keynes, N. 1877. [Recollection of Darwin's honorary LLD degree]. Diary. (CUL-Add.7831.2), transcribed by Kees Rookmaaker. *The Complete Work of Charles Darwin Online* (http://darwin-online.org.uk/).

Keynes, Richard. 2002. *Fossils, finches and Fuegians : Charles Darwin's adventures and discoveries on the Beagle, 1832–1836.* London: HarperCollins.

Larkum, Tony. 2009. *A natural calling: life, letters and diaries of Charles Darwin and William Darwin Fox.* Dordrecht and New York: Springer.

Leedham-Green, E.S. 1996. *A concise history of the University of Cambridge.* Cambridge: Cambridge University Press.

Leighton, W. A. C. 1886. [Recollections of Charles Darwin]. (DAR 112.B94-B98), transcribed by Kees Rookmaaker. *The Complete Work of Charles Darwin Online* (http://darwin-online.org.uk/).

LeMahieu, D. L. 1976. *The mind of William Paley: a philosopher and his age.* Lincoln, Nebraska.

Litchfield, H. E. (ed.) 1915. *Emma Darwin: a century of family letters, 1792–1896.* 2 vols. London: John Murray.

Locke, John. 1690. *An essay concerning human understanding.* London: Basset.

Loggan, David. 1690. *Cantabrigia Illustrata.* Cambridge.

Lowe, Robert. 'Journal kept by H. P. Lowe & R Lowe during 3 months of the summer 1831. at Barmouth. North Wales. Forsitan haec olim meminisse juvabit.' Transcribed from the manuscript by Peter Lucas. http://darwin-online.org.uk/content/frameset?viewtype=text&itemID=NRO-DD.SK.218.1&pageseq=1.

Matric. 11: Books of subscriptions for degrees, Cambridge University Archives, Cambridge University Library.

Moore, J. R. 1994. *The Darwin legend.* Grand Rapids, Michigan: Baker Books.

Paget, G. E. 1882. [Recollections of Darwin, 13 September.] (DAR 112.A86-A91), transcribed by Kees Rookmaaker. *The Complete Work of Charles Darwin Online* (http://darwin-online.org.uk/).

Paley, W. 1802. *Natural theology: or, evidences of the existence and atributes of the deity, collected from the appearances of nature.* London.

Pattison, Andrew. 2009. *The Darwins of Shrewsbury.* Stroud: History Press.

Peile, J. 1900. *Christ's College.* London: Robinson.

Peile, J. (ed.) 1913. *Biographical register of Christ's College 1505-1905 and of the earlier foundation, God's House 1448–1505.* 2 vols. Cambridge: University Press.

Poole, Jo. 2009. The restoration of Darwin's rooms. http://www.christs.cam.ac.uk/college-life/article-jo-poole.

Price, John. nd. [Recollections of Darwin.] (DAR 112.B101-B117), transcribed by Kees Rookmaaker. *The Complete Work of Charles Darwin Online* (http://darwin-online.org.uk/).

Rackham, Harris (ed.) 1939. *Christ's College in former days. Being articles reprinted from the College Magazine.* Cambridge: University Press.

Raverat, Gwen. 1952. *Period piece: A Cambridge childhood.* London: Faber and Faber.

Richmond, Marsha. 2006. The 1909 Darwin celebration: reexamining evolution in the light of Mendel, mutation, and meiosis. *Isis* 973: 447–484.

Rodwell, J. M. 1882. [Recollections of Darwin, 8 July.] (DAR 112.A94-A95), transcribed by Kees Rookmaaker. *The Complete Work of Charles Darwin Online* (http://darwin-online.org.uk/).

Rookmaaker, Kees ed. [Darwin's Beagle diary (1831–1836)]. [English Heritage 88202366] (*The Complete Work of Charles Darwin Online*, http://darwin-online.org.uk/).

Rookmaaker, Kees, 2004. *A Calendar of the historical documents of the University Museum of Zoology, Cambridge 1819–1911.* Cambridge, University Museum of Zoology.

Royal Commission on the Historical Monuments of England. 1959. *An inventory of the historical monuments in the city of Cambridge.* 2 vols. London: Her Majesty's Stationery Office.

Rutherford, H. W. 1908. *Catalogue of the library of Charles Darwin now in the Botany School, Cambridge*. Compiled by H. W. Rutherford, of the University Library; with an Introduction by Francis Darwin. Cambridge: Cambridge University Press.

Secord, James. 1991. Edinburgh Lamarckians: Robert Jameson and Robert E. Grant. *Journal of the History of Biology* 24: 1–18.

Secord, James. 1991. The discovery of a vocation: Darwin's early geology. *British Journal for the History of Science* 24: 133–157.

Shipley, A. E. 1924. *Cambridge cameos*. London: Jonathan Cape.

Shipley, A. E. and Simpson, J.C. (eds.) 1909. *Darwin centenary: the portraits, prints and writings of Charles Robert Darwin, exhibited at Christ's College, Cambridge 1909*. Cambridge: University Press.

Smith, Adam. 1759. *Theory of moral sentiments*. London: A.Millar.

Smith, K.G.V. 1987. Darwin's insects: Charles Darwin's entomological notes. *Bulletin of the British Museum (Natural History) Historical Series* 14(1): 1–143.

Smocovitis, V. B. 1999. The 1959 Darwin Centennial Celebration in America. *Osiris* 14: 274–323.

Snow, C. P. 1972. *The Masters*. Harmondsworth: Penguin Books Ltd.

Steel, Anthony. 1949. *The custom of the room or early wine-books of Christ's College*. Cambridge.

Stephens, J. 1829. *A systematic catalogue of British insects*. London: Baldwin and Cradock.

Stephens, J. 1828–1835. *Illustrations of British entomology; or, A synopsis of indigenous insects: containing their generic and specific distinctions*. 11 vols. London: Baldwin and Cradock.

Thackeray, William Makepeace. 1878. *Etchings by the late William Makepeace Thackeray while at Cambridge: illustrative of university life, etc., etc. Now printed from the original plates*. London: H. Sotheran and Co.

Thomson, Keith. 2009. Young Charles Darwin. New Haven and London: Yale University Press.

Venn, J. A. 1940–1954. *Alumni Cantabrigienses*. Part II, 1752–1900. Cambridge: University Press.

Wallace, A. R. 1889. *Darwinism: an exposition of the theory of natural selection with some of its applications*. London & New York: Macmillan & Co.

Walters, S. M. and Stow, E.A. 2001. *Darwin's Mentor: John Stevens Henslow, 1796–1861*. Cambridge: Cambridge University Press.

Watkins, F. 1887. [Recollections of Darwin, 18 July.] (DAR 112.A111-A114), transcribed by Kees Rookmaaker, *The Complete Work of Charles Darwin Online* (http://darwin-online.org.uk/).

Winstanley, D. A. ed. 1932. *Henry Gunning: Reminiscences of Cambridge*. Cambridge: University Press.

Woram, John. n.d. Portraits in the round: busts of Charles Darwin, http://www.galapagos.to/TEXTS/COUPER.HTM.

Wright, John M.F. 1827. *Alma Mater, or, seven years at the University of Cambridge.* London : Black, Young, and Young.

Wright, Thomas. 1845. *Memorials of Cambridge: a series of views of the colleges, halls, and public buildings.* London: David Bogue.

Wyhe, John van. (ed.) 2002– *The Complete Work of Charles Darwin Online* (http://darwin-online.org.uk).

Wyhe, John van. 2004. *Phrenology and the origins of Victorian scientific naturalism.* Ashgate.

Wyhe, John van. 2007. Mind the gap: Did Darwin avoid publishing his theory for many years? *Notes and Records of the Royal Society* 61: 177–205.

Wyhe, John van. 2008. *Darwin.* London: Andre Deutsch and New York: National Geographic [2009].

Wyhe, John van. 2010 Commemorating Charles Darwin. *Evolutionary review* 1(1)(February): 42–47.

Wyhe, John van. 2013. "my appointment received the sanction of the Admiralty": Why Charles Darwin really was the naturalist on HMS Beagle. *Studies in History and Philosophy of Biological and Biomedical Sciences* 44(3): 316–326.

Wyhe, John van. 2013. *Dispelling the Darkness: Voyage in the Malay Archipelago and the discovery of evolution by Wallace and Darwin.* Singapore: World Scientific Press.

Manuscripts in Christ's College Archives.

Study rents 1741–1782.
Admissions to Christ College (T.1.2).
Admissions 1818–1828 (T.1.4).
Admissions 1815–1852 (T.3.1).
1822–1829 Tutors' Accounts (T.11.26).
1830–1835 Students Bills (T.11.27).
1827–1831 Students Bills (T.11.25).
1828–1831 Study Rent (T.9.5).
1828–1829 Residents Book (T.17.A).
1831–1836 Lecturer's Book (T.8.2).
Darwin — Fox Correspondence, Christ's College Library.

INDEX